THE SOUL OF GENIUS

The 1911 Solvay Conference on Physics. Seated (L-R): W. Nernst, M. Brillouin, E. Solvay, H. Lorentz, E Warburg, J. Perrin, W. Wien, M. Curie, H. Poincaré. Standing (L-R): R. Goldschmidt, M. Planck, H. Rubens, A. Sommerfeld, F. Lindemann, M. de Broglie, M. Knudsen, F. Hasenöhrl, G. Hostelet, E. Herzen, J. H. Jeans, E. Rutherford, H. Kamerlingh Onnes, A. Einstein, P. Langevin.

THE SOUL OF GENIUS

Marie Curie, Albert Einstein, and the Meeting that Changed the Course of Science

JEFFREY ORENS

PEGASUS BOOKS
NEW YORK LONDON

THE SOUL OF GENIUS

Pegasus Books, Ltd.
148 West 37th Street, 13th Floor
New York, NY 10018

First Pegasus Books paperback edition January 2023
First Pegasus Books cloth edition July 2021

Interior design by Maria Fernandez

frontispiece: The 1911 Solvay Conference.
Courtesy of The Solvay Heritage Collection.

Library of Congress Cataloging-in-Publication Data is available.

ISBN: 978-1-63936-217-2

10 9 8 7 6 5 4 3 2 1

Printed in the United States of America
Distributed by Simon & Schuster
www.pegasusbooks.com

To My Wife Deborah
My Life Partner

CONTENTS

INTRODUCTION

In many offices of the Belgian multinational, multibillion-dollar chemical company Solvay S.A., a first impression invariably includes coming face-to-face with what has often been termed "The Most Intelligent Picture Ever Taken." It's a shot of the attendees of the 1927 Solvay Conference on Physics, the fifth in a series that started in 1911 and is ongoing even to this day. The picture is usually in the reception area or in a hallway leading from it to the general offices, covering the better part of a wall, impossible to miss.

Three rows of formally dressed scientists impassively stare out at the observer from the black-and-white photo, twenty-nine people who at first glance appear to be more interested in getting back to their sessions on theoretical physics than posing for the picture. Some faces jump out quickly, others are not so recognizable except to physicists and chemists observing the image. Albert Einstein is easy to spot, of course, in the front row, dead center, the sun surrounded by the lesser planets and stars in the scientific heavens. Marie Curie had also been accorded a seat up front, but slightly to the left, comfortably flanked by two giants of physics a bit less well known but almost as illustrious, Max Planck to the left in the picture and Hendrik Lorentz to the right. Niels Bohr is in the second row, far right, no doubt just prior to the picture having discussed some facet of quantum theory with fellow disciple Max Born next to him.

Seventeen of the twenty-nine at the 1927 meeting had received, or would receive, Nobel Prizes in either physics or chemistry, almost 60 percent of those in the picture. Hence, the well-worn reference to the high IQs of those in the photograph. Noticeably, only one of the photographic subjects is a woman, Marie Curie, by then the only twice-awarded Nobel Prize winner at the time, for her achievements in both physics and chemistry. She had been the sole woman attending these gatherings until 1933, when she was joined at another Solvay Conference by her daughter, soon-to-be Nobel Prize winner Irène Joliot-Curie, as well as the Austrian physicist Lise Meitner.

Looking at the sober faces of Einstein and Curie, Planck and Bohr, knowing the connection between these people and some of their revolutionary contributions to advancements in the world of science, spurred me on to learn more about the meeting behind the picture. Somewhat akin to observing Newton's apple dropping from a tree to the ground in his garden (but on an infinitely less-impactful scale), the intrigue of the image caused me to seek out a more complete understanding of the reasons for these gatherings. It turns out the Fifth Solvay Conference on Physics was significant in its own scientific right, not just for cornering the market on Nobel Prize winners to fill the session but for its famous debates between Einstein and Bohr on quantum theory. These two giants of theoretical physics were embroiled in debate on competing theories concerning probabilities versus certainties. Bohr said that quantum mechanics dictated that probabilities of events were the only way to explain the workings of the subatomic universe while Einstein, one of the original discoverers of quantum theory two decades earlier, was not so sure. Instead, he clung to a more classical view that in the real world, measurements were done with certainty, not probability. In previous comments on probability versus certainty that

Einstein expressed to Bohr's like-minded cohort Max Born, Einstein had famously stated, "I am convinced that [God] is not playing dice."[1]

Wanting to know more about the Solvay Conference in general led me to do some research that opened my eyes to the profound nature of these meetings. Ernest Solvay, whose namesake company now employed me, had been presented with the idea of sponsoring this meeting as part of his philanthropic ventures in support of his enthusiasm for science. The thought was to bring together a core group of the most intelligent scientists, the "best and brightest" of their day, to discuss the most pressing issues of physics. At the time, it was a novel perspective for professional conferences. Previously, much larger gatherings of scientists intermittently occurred to give papers on various subjects, with little discussion taking place following the formal presentations. In 1910, German physicist Walther Nernst suggested a different approach to support discussion and debate concerning classic physics tenets versus the newly developing, subatomic view of the world that was called quantum theory, which was based in part on the idea that light was composed of energy that traveled in waves as well as discrete particles called quanta. The First Solvay Conference was Nernst's brainchild, nurtured by the passion and funding of Ernest Solvay.

Twenty-four individuals congregated in Brussels for the First Solvay Conference on Physics, focused on the theory of radiation (interaction of light and matter) and quanta. Marie Curie was one, as was Albert Einstein, the youngest of the invitees. The results, though inconclusive in solving the puzzles presented at the meeting, generated an excitement within the group to continue these forums for idea exchange, which was formalized into the establishment of a triennial meeting. The most recent Solvay Conference on Physics was in 2017, the 2020 edition being postponed until 2021 due to the COVID-19 pandemic. The 2017 session delved into the physics of living matter: space, time, and information in biology. Fifty-eight scientists participated, nine of whom were women (15 percent). This marked a slow increase in female participation reflecting small but gradual growth in STEM (science, technology, engineering, and mathematics) educational and professional opportunities available for women over the past century, as well as women's career choices moving more in that direction. It represents a determined expansion of female involvement versus Marie Curie's rare desire to become a scientist in the 1890s, a woman fighting the tide of often overwhelming cultural biases of her time.

As I investigated the history of the First Solvay Conference, I discovered the personal stories of its select participants, especially the two brilliant individuals who have made the greatest impression on people across the world when they think of science, Marie Curie and Albert Einstein. Wonderful biographies already exist about these two fascinating subjects, voluminously detailing their backgrounds and accomplishments. I know because I've read a number of them and reference their insights liberally throughout my book. Understanding Curie's and Einstein's experiences, their struggles, and their varied interactions helped me to appreciate the very personal challenges and sacrifices they made in striving to excel in their chosen field. Their meeting at the First Solvay Conference began an enduring friendship that was influential for both.

In researching for this project, examining primary source material proved difficult as the coronavirus appeared in 2020. The pandemic resulted in limited staffing of, and restricted access to, many libraries worldwide. Fortunately, the online availability of a variety of information from both primary and secondary sources about many of the characters and events of this story, especially concerning Curie and Einstein, made the search process more manageable than I could have anticipated.

In this regard, two resources of particular note stand out. The Collected Papers of Albert Einstein is a Princeton University Press/Hebrew University/California Institute of Technology project still undergoing completion but readily available to all on the Internet since 2014. It includes digitized English translations of letters and documents to and from Einstein, currently covering the period from the late 1800s through the 1920s, more than adequate for my purposes. This work contains his writings and correspondence, needless to say a wealth of information about the manner in which Einstein approached the world over this period, whether in formal papers or conversing with scientific peers, personal friends, or the women in his life.

The second item was a find that was as unexpected as it was valuable. In searching for information concerning the liaison between Marie Curie and Paul Langevin, by chance I followed an Internet thread that led me to a finding aid for the papers of physicist Marcel Brillouin, a French contemporary of the two scientists and fellow attendee of the First Solvay Conference. These writings were contained within a collection of the papers of his son, physicist Léon Brillouin, in the American Institute of Physics in College Park, Maryland. Marcel's papers include a file of notes he kept, handwritten in French, describing in great detail the events, as told by Langevin's wife Jeanne to Brillouin's wife Charlotte as well as himself, about the infamous affair between Curie and her husband Paul. Also included is correspondence related to the situation between Brillouin and two fellow physicists, Jean Perrin and Hendrik Lorentz. To my knowledge, this appears to be the first time this Brillouin background information has been publicly utilized in describing the story of the relationship. It presents a stark contrast to the manner in which this star-crossed pairing has been portrayed in past retellings. The journal is much more sympathetic to Langevin as well as unforgiving to Curie, in keeping with how a conservative Frenchman at the time might view the situation. It's an eye-opener, to be sure, comprising the basis for much of chapter eight.

At the same time, digging into the history of the conference itself led me to more fully appreciate the dedication and perseverance of

Ernest Solvay. As he aged from supremely successful businessman to scientific philanthropist, his efforts pointed toward providing scientists like Curie and Einstein with a dynamic forum to share groundbreaking theory and experimentation and to debate the merits of alternative approaches in an open and global manner. The small group setting of that first Solvay Conference encouraged frank discussion in the daily formal sessions as well as at breaks and meals spent in each other's company. The uniqueness of these conferences was noted by the participants, who once invited were excited to return to future gatherings. My good fortune in being able to discuss these conferences with Solvay's great-great-grandson Jean-Marie, who is still very much involved in these gatherings, provided some insightful commentary shared in the epilogue.

Genius indeed comes in many forms, crossing genders, nationalities, and races. For in science, as in other walks of life, once found, if it's carefully nurtured on both a professional and personal level, it has the chance to fully bloom. If not, as many in this book could attest, it can still be exhibited by those strong enough to overcome barriers placed in their paths. Either way, there is little doubt that developing an intelligent, competitive environment drives creativity and innovation, whether in business, academia, or life in general. It's needed to propel our society forward, with our institutions hopefully providing a level playing field and opportunities for success. But just the process of those with unmatched brilliance being brought together to exchange opinions derived from deep thought and experimentation can be supportive as well, sometimes achieving a meeting of the minds and sometimes requiring further exploration and discovery, which can lead to society's ultimate benefit.

CHAPTER ONE

The First Solvay Conference: A New Approach

It was still difficult for Marie Curie to compose herself as she packed up to leave the Hotel Metropole in early November 1911. She had received devastating news, almost on the heels of the elation she had felt only a few days before. Then, it had been a special delivery telegram from the Royal Swedish Academy of Sciences, notifying her of being awarded the Nobel Prize in chemistry. No one had ever won two Nobel Prizes before, her first coming eight years prior as a co-winner of the prize for physics along with her beloved husband Pierre Curie for their research on radioactivity. This second award was for the discovery and purification of the chemical element radium, as well as the discovery of polonium.

Now came the bombshell reports on the morning after the conference concluded, the French press publicizing speculation, for all to read, about her private love letters. Only these were not dated pledges of everlasting adoration for her husband Pierre, who had been dead for five years. Rather, they were much more recent impassioned expressions of devotion to her lover of the past eighteen months, Paul Langevin. He was a distinguished French scientist and mathematician in his own right, a former protégé of her late husband, and, more important, a married man. She was now being labeled a "home-wrecker," and the papers went further, tarring her with a xenophobic brush as a foreigner, being born and raised in Poland, and a Jew, which was a complete fabrication designed to appeal to the deep anti-Semitism still present in much of France and Europe. As the stories began to spread from France to Belgium, where she was attending the First Solvay Conference on "The Theory of Radiation and Quanta" along with Paul, the two had quickly determined it would be best to take a train back to Paris at the first opportunity.

Marie hated to drop everything and scurry back home. It was against her very nature to bow to public pressure and extract herself from the conference, its important mission, and the preeminent scientists who were in attendance. After all, Marie's inherently strong will and principled character had driven her to triumph over so many adversities. She had suffered personal tragedy in her family when she was a child in Warsaw, overcoming the deep despair seared into her soul by these events to achieve her goal of scientific education and discovery. The oppressive Russian rulers of her native Poland had been unable to keep her thirst for knowledge unquenched as she defied their restrictions through clandestine scientific studies with other like-minded young ladies. Finally, she had broken free and come to Paris, where she fashioned a life of study, research, and family with her husband Pierre and two daughters, winning with him a Nobel Prize in the bargain. Hers was a life that had already inspired women around the world.

Now she had gathered in Brussels with others of scientific brilliance at the invitation of the multimillionaire Belgian chemical industrialist Ernest Solvay. He had paid one thousand francs apiece for all twenty-three guests to ensconce themselves in the luxurious Hotel Metropole to debate the very future of how they could reconcile major disruptions of well-accepted theories of physics that were unfolding quickly through scientific experimentation at the atomic level over the past few decades. Solvay had been convinced by German scientist Walther Nernst that this was the surest way to get the undivided attention of these individuals, all men except for Marie, and have them discuss and deliberate over the clash of theories that were threatening the old, long-established Newtonian views of matter and energy. But journalists were now implying that her attendance at the conference with Langevin was just a cover for a pleasure jaunt as part of their illicit relationship. If they did not return immediately, but delayed for even a day, it would certainly provide more grist for the ever-churning press and their insatiable appetite for scandal. She needed to be with her two young girls, and he with his family, to defend themselves as best they could. Marie was determined to fight for her honor against the tabloid innuendo.

Curie's thoughts flashed back to earlier in the year, when she had disturbingly experienced something similar at the hands of the Parisian press. Then, she had been persecuted by the conservative newspapers as unfit to be considered for a chair in physics in the prestigious French Academy of Sciences, in spite of her Nobel award and her well-publicized discoveries of radioactivity and radioactive elements. She had applied for this seat in late 1910, and appeared to have had relatively scant competition. The major exception was Édouard Branly, an engineer who was a pioneer in the field of wireless transmission that eventually supported invention of the radio, and himself a winner of the Nobel Prize for physics in 1909 for contributions to wireless telegraphy. The very idea that a woman could be chosen over a man for a seat in the most exalted scientific assembly in France had the majority of both sexes in the country in an uproar. French writer Julia Daudet and actress

Madame Marthe Régnier commented publicly on the impropriety of Curie's quest for equal treatment by the French scientific community.[1] A Parisian journal, *Excelsior*, the city's illustrated daily paper, had run a front page article on Curie. It was complete with images of Curie from head-on as well as profile viewpoints, much as a wanted criminal would be depicted, attempting to show her "unsavory" foreign and potentially "criminal" characteristics.

But both candidates had their supporters, along with another minor nominee, physicist Marcel Brillouin, and when the final session on the matter took place at the end of January 1911, it was attended by a larger-than-usual throng of public spectators as well as Academy members. A statement about the logistics and attendance at the spectacle said it all: "To prevent incidents, [Academy President] Armand Gautier, who was chairing the session, instructed the ushers to 'allow everybody in . . . except women, of course'!"[2] In the first round of balloting, Branly led Curie by the slimmest of margins, 29–28, with Brillouin a forgotten third with one vote. Since there were 58 total votes and a majority of 30 minimum was needed to win, a second round was required, which Curie lost by two votes, 30–28, never again attempting to join the Academy.[3] Close to seventy years later, the first woman elected to this scientific body as a full member finally broke its long-standing gender barrier in 1979.

Now, Curie was going back to fight the same prejudice and injustice that allowed French public opinion to heap scorn upon her reputation once again. This time for finding love with a married man, separated from his wife, who appeared to be forgiven his sins by much of French society simply because he was a man.

The year before, while Langevin and Curie were nurturing the beginnings of their relationship, Walther Nernst had made contact with Ernest Solvay to propose a special meeting, something that would be unique in the scientific community. Nernst was a strategic planner, and he saw an opportunity to lay the groundwork for a gathering from which he could gain long-term benefit as well as address a burning issue in the scientific community. Classical physics was being challenged by the developing theoretical concept of quantum theory as well as significant experimental evidence to support it, some of it generated by Nernst himself. If key scientific thought leaders could be brought together to agree that the newer theories were credible, Nernst's own work might just be thought worthy of a Nobel Prize.

Certainly, other conferences, congresses, and various formal gatherings had occasionally taken place for scientists, but not on an invitation-only basis for a limited group of the highest-level thinkers. Only one international conference on physics had ever been convened, in 1900 in Paris with over 750 scientists from twenty-four countries.[4] Papers there were presented about work that had been completed, with no ensuing debate conducted on current issues facing the researchers. With only twenty-four proposed participants at the Solvay council, the new format would allow time to present papers to the attendees on important, controversial subjects, followed by energetic discussion and perhaps even development of alternative solutions to some of the issues. Nernst suggested that by offering to host these individuals in an all-expense-paid manner, Solvay could ensure the greatest probability of full attendance. And, for good measure, he noted that it would be an ideal gathering in which Solvay, who had his own particular views of physics and science in general, could address the assembly.

Solvay was agreeable to most everything proposed. But he wanted further discussion on the composition of the list of attendees such that it was more evenly balanced across the main countries that would be sending participants. He also wanted to make sure they had selected an appropriate moderator. Then he let Nernst draft an invitation that Solvay signed and sent in June 1911 to the chosen participants for the conference, set to begin in late October. Twelve of the invitees were also requested to prepare and deliver reports on specific subjects in which each was an expert. The papers were to be sent to all the attendees beforehand so that they would have time to read and reflect on the contents before the actual conference. The invitation stated, in the scientific vernacular of physics, the overriding issue that urgently required their attention, "We are at the moment in the midst of new developments regarding the principles on which classical molecular and kinetic theory of matter has been based. First of all, the logical development of this theory leads to a radiation formula whose validity is incompatible with all empirical results; furthermore, from this same theory also follow the propositions on specific heats . . . that are likewise refuted by numerous measurements."[5] More plainly put, this amounted to a dire warning for the scientific world: We are on the verge of tearing down and replacing a major portion of the foundation of our own long-standing scientific house, and we can't all adequately agree on an explanation as to why we are doing it.

Much of this schism in scientific theory was caused by the ideas generated by the brilliance of Albert Einstein. At thirty-two, he was the youngest scientist invited to attend the conference. A German-born Swiss resident then a university professor in Prague, Einstein at the time of Solvay's meeting was beginning to become the most inventive physicist, and ultimately the most famous scientist, of all time. In 1905 he had authored four papers that eventually would stand the scientific world on its head. To the general public, perhaps the most famous of these was his treatise on the special theory of relativity, $E=mc^2$ (energy equals mass of a body multiplied by the speed of light squared), which stated the direct relationship of energy to mass and theoretically provided the basis for understanding the awesome energy contained in the mass of a particle as tiny as the atom. Somewhat remarkably, due to political infighting in the scientific community, instead of being awarded a Nobel Prize for this insight, he actually received the prize in 1921 for work in theoretical physics related to explaining the photoelectric effect. This was an important concept that he suggested in another paper he authored in 1905, postulating that light sometimes behaved like a wave of energy and sometimes behaved like discrete particles of energy, or quanta. At the Solvay conference, this radical departure from classic physics' interpretation of light as strictly a wave of energy was one of the key pillars of scientific theory that was under attack, and Einstein was leading the charge.

Altogether, of the twenty-four attendees at the First Solvay Conference, nine had won, or were to win, a Nobel Prize in either physics or chemistry, almost 40 percent of this illustrious group. Never had there been a concentrated gathering of such intellectual capacity. Reading their names was like listing members of a scientific Hall of Fame. Hendrik Lorentz was a Dutchman who was asked to chair the event because of his renowned intelligence, amiability, and mastery of multiple languages. He had been awarded the 1902 Nobel Prize in physics with a pupil for their theory on electromagnetic radiation. Max Planck was a Berliner whose revolutionary concepts concerning energy and light were the basis for the development of quantum theory even before Einstein, and for which he was awarded the 1918 Nobel Prize in physics. Jean Baptiste Perrin, a Frenchman, gained the 1926 Nobel Prize in physics for the experimental confirmation of the atomic nature of matter. He and his wife were close friends with the Curies. Nernst, the German chemist, spent a lifetime investigating thermodynamics, or the relationship among temperature, work, heat, and energy. Much of his experimentation supported the newly developing quantum theory, of which the First Solvay Conference was to spend much of its time debating and subsequently agreeing on, bolstering Nernst's bona fides for which he finally did receive the 1920 Nobel Prize in chemistry. Besides these four, plus Curie and Einstein, the other Nobel Prize–winning scientists in attendance included Ernest Rutherford (1908 in chemistry), Wilhelm Wien (1911 in physics), and Heike Kamerlingh Onnes (1913 in physics).

Genius was certainly a term that applied to the participants, but not only because of each attendee's inherent intellectual abilities. From Thomas Edison's more practical definition of genius, "one percent inspiration and ninety-nine percent perspiration," the word as applied to this assembly signified a varying combination of insightful idea generation in the fields of physics and chemistry combined with tireless, dedicated efforts to prove these ideas. Often, these experiments took days, weeks, or months in order to develop workable hypotheses and then show incontrovertible proof of the nature of these thoughts. In some cases, it took years to develop just the right conditions and employ specially devised equipment to prove a theory.

That was the situation for Einstein's general theory of relativity, famous for its leading to the expectation that starlight would be bent, however imperceptibly, when in close proximity to a very large mass, like a star in the solar system. Classical Newtonian theories of gravitation also predicted this phenomenon, but not nearly to the degree foretold by Einstein's theoretical work. Proposed in 1915, it wasn't until a total solar eclipse took place four years later that this hypothesis could be proved, and astronomer Arthur Eddington's work showed just that with photographic evidence produced during the eclipse. The world was astounded, the *New York Times* headline of November 10, 1919, reading, Lights All Askew in the Heavens—Men of Science More or Less Agog over Results of Eclipse Observations—Einstein Theory Triumphs.[6] People were utterly amazed as they tried to comprehend how light could be bent, let alone by as much as Einstein had predicted, but science now had tangible evidence to prove it could be done.

Similarly, it took years of experimentation for Marie and Pierre Curie to prove their hypothesis that radioactivity was a property inherent to many heavy elements, not caused by chemical reactions between elements and compounds, but actually a product of atomic decay of the elements themselves. Then, it took even longer to develop and implement the processes to purify the element of radium for identification, for which Marie Curie herself was to win her second Nobel Prize. Instrumental to all of this was the funding needed to carry out the experimentation to demonstrate proof of these concepts.

As scientific discovery progressed during the late 1800s, becoming more complex, especially related to investigations at the atomic level, invention of experimental methods and specific tools of observation were required. Repetition of these experiments for reproducibility of results was just as important in showing that a theory could be reliably reduced to practice. Having an adequate laboratory to make all this happen, one with enough space for conducting experiments, enough of the right kind of equipment, and enough trained personnel to proceed efficiently and effectively along the way was almost as much a goal of each experimental scientist as were the generation of ideas that needed exploring.

As such, monetary grants were increasingly sought in order to finance the mission of science, to understand the natural world through observation and experimentation. As the 19th century came to a close, it became exceedingly difficult for isolated individuals of intelligence and independent wealth to serendipitously delve into the world's natural mysteries in search of answers. In the arts, the sponsorship of royal or wealthy patrons still allowed many a genius to be creative and productive in a leisurely manner, whether painting a portrait or composing a symphony. In the sciences, many European countries had associations that pursued greater understanding of the physical world through granting of monies to individuals or teams deemed worthy by association members of support for exploration and discovery. The Academy of Sciences in France was one such organization, reviewing scientific papers and funding specific investigations. Major cities in many of these countries had universities that sought professors who could promote in-depth investigation for their students and themselves. A number of the more prestigious universities paid well, and offered their professors laboratories that had state-of-the-art equipment and facilities. Germany led the way, with centers of higher learning that strongly supported scientific research across the country, including in Berlin, Munich, and Leipzig. France had the Sorbonne. England's Cambridge and Oxford were scientific research focal points, as was Copenhagen's University in Denmark, just to name a few. Competition became more intense in the scientific community to secure professorships at these top universities, with their generous salaries and laboratory resources.

Then, in 1896, the Swedish-born chemist, engineer, international businessman, and philanthropist Alfred Nobel died. He had spent much of his life building a fortune based on the explosive power of nitroglycerine. Nobel learned how to control its volatile destructiveness by blending it with additives that made it more stable in processing and handling. The result was the patented material dynamite, a product that revolutionized the world for both better and worse. For better by providing a product that could be used simply, easily, and cheaply to blast tunnels, mines, and canals around the world. This paved the way for changes that assisted the digging for mineral resources as well as providing the means to build infrastructure that allowed people and products to move across the globe in a more efficient manner. For worse in that dynamite itself had the power of immense human carnage when used to make explosive munitions.

When Nobel died at the close of the 19th century, he was concerned about this deadly legacy of dynamite and thus determined to leave one more lasting mark on the world. His will dictated the use of the vast majority of his fortune to fund highly prestigious, generous monetary awards. It stipulated that this capital "is to constitute a fund, the interest on which is to be distributed annually as prizes to those who, during the preceding year, have conferred the greatest benefit to humankind . . . divided into five equal parts . . . in the field of physics . . . chemical discovery or improvement . . . discovery [in] physiology or medicine . . . in the field of literature . . . and advancing fellowship among nations, the abolishment or reduction of standing armies, and the establishment and promotion of peace congresses."[7] The Nobel Prizes were born, the richest grants ever established for use by the ingenious and industrious few who were judged as having scaled the heights of scientific discovery or furthered the cause of peace.

Beginning in 1901, the prizes for physics and chemistry were awarded by the Royal Swedish Academy of Sciences in ceremonies of great pomp and circumstance. In addition to a gold medal and diploma presented to the recipient by the King of Sweden, a cash amount followed shortly thereafter. In 1903, when Marie Curie and her husband Pierre received the award for physics, it was also given to a third recipient, physicist Henri Becquerel, whose experiments on uranium had inspired the Curies to do their groundbreaking study on radioactivity. The cash award was a significant sum, 70,000 francs, worth about $18,000 in 1903 and close to $500,000 in 2019 dollars. With multiple winners for physics, there were two 70,000 franc awards provided that year: one to Becquerel and one to the Curies as a pair.[8] Many in the scientific community felt the Curie award was actually being bestowed upon Pierre, who they viewed as having only received assistance from his wife Marie rather than both being equal contributors. In fact, the official nominating letter for the award had originally spoken of only Becquerel and Pierre Curie, entirely omitting Marie from recognition. Only after Pierre complained vociferously was it agreed that Marie would be added to the nomination, eventually becoming part of the award.[9]

What to do with 70,000 francs? Earlier that year, the Curies had been invited to attend a session of the Royal Society of London, the oldest scientific association in the world. Pierre gave a talk about the discovery of radioactivity, with Marie in the audience as the first woman ever allowed to attend a Royal Society meeting.[10] Over the next few days, they were honored at banquets and receptions hosted by the cream of London society. At one such dinner, after admiring the brilliantly sparkling jewels adorning their hosts and guests, the plainly dressed Marie remarked to Pierre as they retired for the evening that she had spent much of the dinner speculating on how many research facilities could be built with each woman's necklace, finally thinking that, with all the precious stones in the room, it was almost too great a number to calculate.[11] Certainly, though both Curies had been content to live the spartan life of scientists dedicated to their investigative efforts, the sudden windfall of 70,000 francs, courtesy of a Nobel Prize, was the ultimate funding achievement. In Marie's view, ever practical and supportive of her husband, it was truly a godsend, "a unique chance of releasing Pierre from his long hours of teaching, of saving his health!"[12] Not a thought of how it might improve *her* life was mentioned in her correspondence or comments to others about their good fortune.

Philanthropy of this magnitude was not unheard-of in the scientific community. The trend had been developing since the early 1870s, mainly in the form of magnanimous giving to support university scientific studies. The Clarendon physics labs at Oxford as well as the Cavendish research center at Cambridge were early examples, as was Andrew Carnegie's financing of the California Institute of Technology in 1891.[13] However, what was rare in this mix was the intense personal interest that a particular philanthropist, Ernest

Solvay, had taken in the sciences, specifically physics, chemistry, physiology, and sociology. Pursuit of these subjects became the focal point of his life's study and charitable giving in the early 1900s. This, even though due to ill health he did not receive university instruction but rather was a self-taught scientist. Interestingly, Alfred Nobel, though receiving training as a chemist and engineer, had always experienced poor health as well and never attended a university to receive a degree. Other similarities between Solvay and Nobel were quite noticeable, from their amassing fortunes based on their own novel chemical processes and products to their sharing in the value of scientific investigation and discovery and desire to generously reward these explorations.[14] Here were two philanthropists who ardently desired to support the sciences but in slightly different, but significant, ways. Nobel for demonstrated achievements, Solvay in support of exploration, investigation, and development of potential future solutions to scientific phenomena.

By the 1890s, Solvay had become spectacularly wealthy through his practical invention of a more efficient, and thus less costly, process to manufacture soda ash (the compound sodium carbonate) for use in making glass and soaps. But he had always sought to spend more time concentrating his efforts on theoretical science, intuitively convinced that there was a unifying principle or law waiting to be discovered that could unite physics and chemistry with physiology and sociology, an "all-embracing Law" that could connect them all.[15] He spent much of his effort and money at the turn of the 19th century searching for this unifying theory, but time wore on and his hypotheses in this regard were not accepted by the scientific community. It was then that he found his real mission, and talent, was to bring the most brilliant scientific minds of his age together to share insights and speculation and find solutions for the world's most difficult issues affecting these four scientific disciplines.

When Walther Nernst saw that his own novel experimental results were moving him toward agreement with Max Planck's and Albert Einstein's views on quantum theory, he moved to take advantage of Solvay's willingness to financially back a conference of the highest level to discuss and hopefully agree on this radical quantum departure from classical physics views on light and energy. Through a mutual acquaintance, Belgian University of Brussels professor and chemist Robert Goldschmidt, Nernst approached Solvay with what he knew would be a proposition too good for Solvay to resist. It was time for the selected elite brainpower of the European physics community to assemble to have an open discussion concerning the earth-shattering possibilities of quantum theory as it was unfolding in real time. Hopefully, this would allow them to come to some sort of consensus concerning this most vexing of problems to reconcile with classical Newtonian physics gospel. Nernst was anxious to have scientific agreement on quantum theory to justify his own findings. He felt it would be appropriate for someone like Solvay, who desired nothing more than the supercharged networking opportunity this event would provide, to spearhead the effort and sponsor the select list of attendees in Solvay's hometown of Brussels.[16] Besides, Solvay felt the very real need to address this scientific conundrum head-on. His own great-great-grandson Jean-Marie noted in a current-day article on the subject, "Ernest Solvay himself understood the reality of this crisis, and wanted to participate in its resolution."[17]

Einstein readily agreed to attend, and responded positively to Nernst on the "Solvay" invitation, "I accept your invitation to the meeting in Brussels with pleasure, and will gladly write the report you had in mind for me. I find the whole undertaking extremely attractive, and there is little doubt in my mind that you are at its heart and soul."[18] They had only just met the previous year when Nernst had visited Einstein in Zurich. His experimental work was coming closer to Einstein's theoretical quantum views, and Nernst wanted to evaluate the young scientist in person. He wrote of the encounter, "I visited Prof. Einstein in Zurich. It was for me an extremely stimulating and interesting meeting. I believe that, as regards the development of physics, we can be very happy to have such an original young thinker . . . Einstein's 'quantum hypothesis' is probably among the most remarkable thought [constructions] ever."[19]

Due to Einstein's revolutionary yet intriguing theories, he had begun to steadily climb from obscurity. The last few years had been a blur of activity and career change for him. In 1909, Einstein left his position as a Swiss patent examiner. Almost simultaneously, he was invited to receive an honorary doctorate at the University of Geneva for his groundbreaking theoretical physics papers and their influence on Charles Guye, a physics professor at the university, who recommended Einstein for the honor.[20] In July, the university was throwing itself a party, celebrating the 350th anniversary of its founding by the austere and righteous reformer John Calvin. Coincidentally, among the over 100 recipients of the awarded doctorates were both Marie Curie and Ernest Solvay; the three may have made their initial acquaintance at this event. At the evening's grand feast, Einstein commented in his discussions with a Geneva resident in attendance, "Do you know what Calvin would have done if he were still here? . . . He would have erected a large pyre and had us all burned because of sinful gluttony."[21] No further conversation followed between the two, as Einstein's characteristic witty observations, often humorous and always incisive, were rarely filtered.

At the same time, Einstein continued to experience increasing strain in his marriage. He had met his wife, Mileva Marić, when they were both students at the Zurich Polytechnic, a technical college. Mileva was herself a solid mathematician and physicist, a few years older than Albert, but they seemed to intuitively connect. He did not consider her attractive in a classical sense, but rather for what she represented: a person who understood the way he thought, who could discuss ideas of the physical world and beyond, who could perhaps share his wildest notions about the universe. But, as their time together wore on, even after they had begun a family with two boys that they both seemed to truly care for, Einstein was to become more distant.

Later that year, in October, Einstein started a position as an associate professor of physics at the University of Zurich. Before he settled into the job, he gave a speech at a physics conference that threw down the gauntlet to classical theory. This was the first time Einstein had ever been asked to deliver a paper to a gathering of physicists. He gave a presentation before more than one hundred scientists gathered in the picturesque town of Salzburg, high in the Austrian lake district, at the 81st Congress of German Natural Scientists and Physicians. In a topic that pitted classical physics against the new quantum theory, Einstein expounded on why light could be a wave as well as a discrete packet of energy. Those in attendance were intrigued by the theory, and debated it vigorously. Wolfgang Pauli, a future groundbreaker himself in quantum theory, stated the speech was "one of the landmarks in the development of theoretical physics."[22] Others were conflicted about this definitive statement of the nature of light, feeling it went too far. They preferred to explain light classically as a wave of energy, grudgingly admitting that it might take the form of a particle or packet of energy only in certain, very specific, instances. Battle lines were being drawn.

But even as he prepared for the Solvay conference in the early fall of 1911, Einstein was becoming somewhat skeptical of the importance of the gathering. He confided in a letter to close friend Michele Besso that the preparations for the meeting might be more trouble than they were worth.[23] A month later, with the conference only weeks away, he again gave Besso a rather dim view of the whole affair, anxious to be free of the demands being placed on his time.[24] These comments appear to be the product of an overworked professor concerning a conference where he had been assigned yet another burden, the preparation of a paper, in the midst of everything else on his plate. It seems he was giving little thought to his role as a, if not *the*, main player in the drama of theoretical physics to be played out both in Brussels and for years to come.

Yet, for Einstein, the First Solvay Conference was no less than his scientific "debutante ball." He had individually met a number of the most learned scientific personalities by the time of this event, including Lorentz, Nernst, and Planck. In Salzburg, he had been exposed to key German-speaking physicists. But the handpicked gathering of international geniuses from across Europe that attended the Brussels session was uniformly at another level. He was only thirty-two and just an associate professor with an honorary doctorate from the University of Geneva. While the four major papers that had put him on the scientific map had been published in 1905, a number of the Solvay Conference attendees had not been exposed to them and their radical thoughts, even by 1911. Some had not spent sufficient time contemplating the ramifications of the ideas contained in the papers, which was precisely what the conference was intended to allow them to do—digest ideas, discuss the merits, decide on a path forward. The very process of presenting his paper would allow those in attendance to see how his mind worked and how he expressed his thoughts. Was he straightforward in his explanations, was he convincing in his arguments, did he have a plan of action to take things to the next level? During discussions that followed his presentation, as well as those of others, did he question in a pragmatic manner, did he offer alternative explanations of his own design, did he build on the ideas of others or proceed to tear them down?

First impressions inevitably are lasting ones, and in Einstein's case they were deservedly quite favorable. Planck and Nernst had met him before, but not in a setting like this. They were deeply impressed by his discussion of quantum theory, so much so that eighteen months after the conference they were heartily recommending him to the Prussian Academy of Sciences for full membership and for a research professorship award.[25] Marie Curie and Henri Poincaré of the French contingent at the council both found him of the highest caliber as a thinker. In a recommendation letter Curie was asked to provide by the university when Einstein applied for a professorship at the Swiss Federal Institute of Technology in Zurich a few months after the conference, she replied in part, "In Brussels, where I attended a scientific conference in which Mr. Einstein took part, I was able to appreciate the clarity of his mind, the vastness of his documentation and the profundity of his knowledge."[26]

Meanwhile, Einstein was forming opinions about his conference-mates. Shortly after the gathering, in a letter to his friend Heinrich Zangger, he provided his reflections and heartfelt feelings about key participants. "H. A. Lorentz is a marvel of intelligence and tact. He is a living work of art! In my opinion he was the most intelligent among the [theoretical physicists] present. Poin[c]aré was simply negative in general, and, all his acumen notwithstanding, he showed little grasp of the situation. Planck stuck stubbornly to some undoubtedly wrong preconceived opinions . . ."[27]

And, of course, he wrote of Marie Curie and Paul Langevin.

"I spent much time with Perrin, Langevin, and Madame Curie, and became quite enchanted with these people . . . The horror story that was peddled in the newspapers is nonsense. It has been known for quite some time that Langevin wants to get divorced. If he loves Mme Curie & she loves him, they do not have to run off, because they have plenty of opportunities to meet in Paris . . . Also, I do not believe that Mme Curie is power-hungry or hungry for whatever. She is an unpretentious, honest person with more than her fill of responsibilities and burdens. She has a sparkling intelligence, but despite her passionate nature she is not attractive enough to represent a danger to anyone."[28]

The conference that Nernst had envisioned and Solvay had decided to financially support was indeed about to bring a remarkable cast of characters together to generate discussion concerning scientific topics of utmost importance related to the physical world and how it really worked. On another equally significant level, it would show that genius is, after all, only human.

CHAPTER TWO

The Intensity of Marie Curie

During the 15th century, the European world was awakening from a millennium of intellectual stagnation, resuscitated from its coma by a rebirth of ideas, learning, and humanistic thought known as the Renaissance. Beginning in the city-states of northern Italy, especially Florence as supported by the patronage of the ruling Medici family, the movement gained momentum and spread across the continent, mainly north into France, Germany, and England. But this Italian influence also moved through Hungary and reached the kingdom of Poland by the late 1400s. At that time, Poland was the dominant partner in a union with Lithuania, a powerful combination with a dynasty of rulers that was not only one of the most respected in Europe, but strong enough to defeat the German Teutonic Order of Knights in 1410 to keep the kingdom free from Prussian control for centuries to come.[1]

In the first decade of the 1500s, the country's royalty welcomed an influx of Italian artists and craftsmen to transform much of the dynasty's fire-damaged castle from medieval to Renaissance style.[2] The kingdom's oldest seat of higher learning, Cracow University, already boasting a number of respected teachers from outside of Poland, continued to attract talent from across Europe. Numerous merchants settled in the region, making investments and stimulating the economy. The future king, Sigismund I, was tutored by an Italian writer and historian who educated him in the humanism of the Renaissance.[3] He ascended to the throne in 1506, and by the time he took as his second wife a princess from the Duchy of Milan in 1518 he could not have been more supportive of the movement, ushering in what would be considered the "Golden Age" of Polish culture and power, art and architecture.

Against this backdrop, the Kopernik clan was of the growing merchant class. A branch of the family located north of Cracow in Torun, a port town of Germanic origin on the Vistula River, which was now under Polish rule. After his father died in 1483, ten-year-old Mikolaj Kopernik of Torun was taken under the wing of his uncle, a wealthy merchant who became a bishop and helped facilitate the boy's education at schools of great renown. Beginning in 1491, he attended the University of Cracow, where he was a student of the celebrated astronomer Albert Brudzewski. An impassioned teacher who supported independent thought, he "aroused his students' interests . . . The discussions he provoked were often carried into the street, not infrequently ending in fist fights among students holding rival opinions."[4] At the same time, he instructed his pupils in the use of some essential tools of the science of astronomy, astrolabes and triquetrums, mechanical devices that were on the leading edge of what passed for technology of the day. They were used to identify stars or planets by determining the altitude above the horizon for these heavenly bodies.

Over the ensuing decade, Kopernik appeared to turn learning into a profession. He studied at the University of Bologna, one of the leading Italian schools of higher education, where he applied himself to mathematics and religious law. Then, he was off to the other great Italian university, located in Padua, taking courses in medicine. Finally, his extensive Renaissance education was completed by his early thirties with his being awarded a doctorate in religious law from the University of Ferrara. During this time, his uncle had obtained for Mikolaj a lifetime administrative position in the Church, eliminating any future financial worries.[5]

Free to devote much of his time to a love of astronomy, Mikolaj Kopernik, or Nicolaus Copernicus as he became known to the world, discreetly developed his heliocentric theory of the solar system, in the process helping to spark science's reawakening in Europe as a modern discipline. In addition to conducting his church administrative duties and acting as a physician to his uncle, he spent much of the remainder of his life on his observations of the heavens and piecing together detailed mathematical proofs that supported his radical hypothesis that Earth and the rest of the planets revolved around the sun. His remarkable treatise on the subject, "Concerning the Revolutions of the Heavenly Spheres," was finally published while he lay on his deathbed in 1543. The Church didn't strongly oppose the manuscript, at least at first. But, in time, it was viewed as heretical by both the Church as well as the Reformation, as much for its disruption of astronomy's status quo as for its contradiction of religious doctrine.

Although his concept had roots buried deeply in known postulates over fifteen hundred years old, during the time of the Roman Empire the Greek astronomer Ptolemy would successfully institutionalize the geocentric theory of the sun and planets revolving around a stationary Earth, in large part supported by the blessings that followed from the Catholic Church. After the publication of the Copernicus model, slowly over the next few centuries its tenets overcame scholarly skepticism and Church suspicion to become accepted as a given of astronomy. Many of Copernicus's views and calculations were subsequently disproven, such as the planets revolving around the sun along circular paths rather than the elliptical orbits familiar today. But, as the English historian of astronomy A. M. Clerke described it, "by the adoption of [this new system], man's outlook on the universe was fundamentally changed."[6]

Four hundred years after Copernicus entered Cracow University, Maria Skłodowska, a twenty-four-year-old Polish expatriate, would realize a lifelong dream and enroll in the Sorbonne in Paris to begin university studies that would transform her life and bring her global fame as Marie Curie. The astounding achievement of Copernicus had been to examine the vast magnitude of the heavens and propose a new approach to the basic principles of its workings. Now Curie was to point her focus in the other direction, toward the infinitesimally minute world of atomic matter and uncover one of its most curious properties, radioactivity, in the process discovering the elements polonium and radium.

Copernicus and Curie were incomparable minds of their Polish homeland. The privileged life of Copernicus, based on his kinship to a wealthy, powerful bishop and royal advisor who arranged for a promising future, and the very fact that he was a man, allowed him easy access to the best of Europe's centers of learning and their available tools of scientific investigation. As well, courtesy of his uncle's connections, Copernicus was given a sinecure that provided ample resources to support himself while he studied and then formulated his theories at leisure. Immersed within the clergy's hierarchy, on a frigid winter's eve Copernicus could be found tucked away in the residential study in his cathedral at Frombork, warming by the fire while enjoying the remains of his pig's knuckle and cabbage stew, studying a celestial globe on a stand next to his chair. An astrolabe lay on his desk along with unfurled maps of the constellations as he contemplated thoughts of the motion of the planets about the sun.

Curie's ambitious journey from Poland to the Sorbonne and beyond was markedly different, a continuously uphill struggle of concentrated fortitude to overcome two main obstacles. First and foremost, that she was a woman. As such, she was barred from enrolling in her own country's centers for higher learning, four hundred years after Copernicus had attended Cracow University. To complicate matters, she had precious little money. In her early days at the Sorbonne, this impediment forced her to live in a barely heated, barren, sixth floor garret. After paying rent, often she had only a few francs left for food, mainly subsisting on scraps and her love of study. Thus, her efforts in pursuit of her education were physically grueling as well as mentally taxing. She had lived with xenophobia as a constant companion from her earliest days as a child in Poland under Russian rule, where these conquerors viewed the Poles with derision. Growing up, she was even forbidden to use her native tongue in the classroom. Yet, despite all the obstacles that were placed in her way, she was to make one of the most astounding scientific discoveries of all time.

Maria was one of five children born to the Skłodowskis in Russian-controlled Poland. Generations earlier, both of her parents' families had been of minor nobility, landowners of modest wealth. Then the Russians took control of much of the country during its partitioning with Prussia and Austria in the late 1700s and again after the Napoleonic wars, stripping away most native Polish landowners' property and leaving nothing but tattered memories.

Both of Maria's parents were intelligent, practical people, employing their brilliance in educational careers. Władysław was a mathematics and physics professor at a boys' high school, or gymnasium, while Bronislawa was the director of a girls' academy. Under the cruelty of their Russian conquerors, they were required to conduct their classes speaking only Russian, with the elimination of any vestige of Polish history, culture, or language. The tsar's presence was an all-seeing eye, spies everywhere, ready at an instant's notice to dutifully report and punish those who did not comply, always watchful to extinguish any signs of Polish nationalistic embers that still might be smoldering inside patriotic hearts. As Marie Curie later commented in her autobiographical notes, "Russian professors, who, being hostile to the Polish nation, treated their pupils as enemies."[7] This resulting tension accompanied every interaction of Pole and Russian, subservience expected of a Pole when in contact with a Russian occupier.

In this repressive environment, Maria, nicknamed Manya and the youngest of the children, lived with her family in Warsaw. Władysław was the dignified professor, a kindhearted man who doted on his four daughters, one son, and his irreplaceable Bronislawa. She was a caring yet somewhat reserved presence in their lives, having developed tuberculosis and with it an accompanying fear of transmitting this dreaded disease to her young children. The result was that the girls especially were deprived of the natural mother-child contact that would have accompanied a loving family relationship.

Manya, along with her brother and sisters, worshipped both parents. Their father was a fount of knowledge that came from his mastery of multiple languages as well as the physics and mathematical curricula that he meticulously taught to his students and, to a lesser extent, his children. The mother was a blend of exceptional characteristics: attractive, culturally refined, and supportive in nature. Yet they could only gaze upon her lovingly from afar, the constant hacking, consumptive cough always in the background, a terse warning of what might befall them if they came too close.

The superior intellects of the parents were inevitably passed along to their offspring. With Manya, the result was even more than could have been expected. Reading at four, memorizing long passages for later recitation, quick to grasp mathematics in all its variations, she was consistently the student that teachers called on when in need of the right answer to any of their questions. Yet, she was shy and reticent about her obvious brainpower, not wanting to embarrass others with her knowledge and talents while inwardly taking great pride in what she could accomplish with her intelligence.

Accompanying her brilliance was a capacity to focus, to block out surrounding background noise and distractions as her wonderful mind carefully investigated, digested, and absorbed the object of her studies. Manya had a driving intelligence that encompassed much more than the ability to memorize vast quantities of information. Her intensity demanded that she attempt to dig more deeply into the meaning behind the words she was reading and the concepts they described in order to arrive at understandings of the very principles that supported the thoughts displayed before her. This intellectual passion would eventually form the basis of her future successes, even when faced with, at times, overwhelming hardships, a characteristic that distinguished her from others of her age.

Generally, the siblings made a joyous group of playmates. Always eager to learn from their father, they spent hours transfixed as he periodically read historical or literary books aloud, effortlessly translating French, German, or English texts into Polish as he proceeded.[8] Professor Skłodowski knew many languages, but insisted on proudly reading each volume in his native tongue as he explored these stories with his children at home, away from prying Russian eyes. He was a fierce nationalist, detested the Russian invaders, and diligently raised his children in that mold.

Then, little Manya's world was ravaged as death visited the family. First, the oldest of the four girls, Zosia, in her early teens, died in a typhus epidemic that swept through Warsaw in 1874. A few years later, tuberculosis finally claimed the life of her beloved mother while still in her forties. Manya was only ten when her mother passed away. She described her mother's death as "the first great sorrow of my life . . . Her influence over me was extraordinary, for in me the natural love of the little girl for her mother was united with a passionate admiration."[9]

She deeply mourned the loss of each, internalizing these deaths with a reaction of "profound depression," a condition that would manifest itself periodically in the future when misfortune struck.[10] Although she carried on with life's duties after each tragedy, her vitality was sorely taxed as she attended school and did her chores. For weeks on end, an overwhelming urge to steal herself away and endure tearful spasms of grief was the only manner in which she could cope. Especially with the death of her mother, someone she idolized but who could never be physically approached, Manya felt helpless to express the cavernous emptiness and loss that permeated her being. She railed at a world that could allow this to happen to such wonderful people as her mother and sister, beginning the process of abandoning her faith in God. As she grew into adulthood, she carried this depressive tendency and loss of religious belief with her, overwhelmed with its wrenching power to incapacitate her on several occasions of personal tragedy.

Still, Manya and her three surviving siblings strove to make their father proud as they exhibited their superior minds in their respective schooling. At fifteen, she graduated first in her class, winning a gold medal for her efforts, brought home to her beaming father just as her brother Jozef and another sister, Bronya, had done previously. Yet, this achievement seemed an empty triumph, as Zosia and her mother were not there to share in the accomplishment. Manya was still depressed and her father suggested that time spent with relatives, away from Warsaw and its vivid memories of family loss, could hold a possible cure for her condition.

So it was decided, and a year was devoted to this endeavor, with encouraging results. This break from the grind of continuous study and painful memories, a time to spend with a network of uncles, aunts, and cousins in a series of households away from the city and deep into the Polish countryside, was a soothing tonic for Manya's aching soul. The days passed with mornings rising to birdsong or the eerie, rhythmic chant of a barn owl as the sun crept past the bedroom curtains, nothing planned for the day beyond picking wild strawberries with her cousins after a big farm-style breakfast of sausage, eggs, and thick slices of just-baked black bread spread with butter fresh from the churn. She learned how to ride and fish, attended parties with family and neighbors, sometimes dancing until dawn, all reawakening in her the simple joys of being alive.

While at one uncle's home, Manya became enchanted with the Carpathian mountains. Nestled in a valley with the Tatra range of high peaks for a background, Warsaw seemed a fleeting memory. As her daughter Eve was to describe her mother years later, "she was struck with wondering stupor at the snowy glittering summits and the stiff black firs."[11] Finally, Manya had to return to her own home, but not without bringing a piece of the Carpathians firmly implanted in a corner of her heart. She had become a lifelong lover of the outdoors, of meandering footpaths that led to hidden treasures; here, a mirrored pond surrounded by a meadow of brilliantly colored wildflowers; there, a rocky outcropping above the tree line from which to catch a breath-taking view of the mountainous horizon. In the process, she also came home a young woman; at sixteen, a child no longer.

Over the years, the family's economic fortunes had deteriorated. Professor Skłodowski had previously lost much of his small savings in an ill-advised investment. Then, he had been demoted in his position at school by a vengeful Russian superior, forcing the family's move to a succession of smaller lodgings while taking in boarders to augment their now meager earnings. It had been from one of these boarders that Zosia had contracted the deadly typhus virus. Manya, along with her sister Bronya, began to tutor others in order to supplement their father's scant income. Although they were paid little, each ruble was helpful.

Jozef, Bronya, and Manya all thought of higher education as their calling, but this required funding. For Jozef, who wanted to become a medical doctor, attending Warsaw University was the next logical step, which he pursued with the support of their father, poor though he was. Bronya wanted to become a physician as well, but there was no university in Poland that allowed women. Similarly, Manya wanted to take a next educational step, but the same question arose for her as a woman: where?

The Skłodowska sisters searched for an answer, finding it in the workings of a recently formed, clandestine organization in Warsaw, colloquially known as the "Floating" or "Flying" University. This was a series of classes offered in the sciences, arts, and natural history, loosely structured by dedicated Polish women instructors who wished to provide other like-minded women a chance at higher education that could otherwise not be obtained by them in Poland. The classes met in relative secret, "floating" between teachers' homes or behind closed doors of various private institutions, so as not to arouse the ire of the occupying Russians, who opposed this venture for two reasons. First, they were against women receiving higher education in general. Even more important, they were suspicious of these classes as being a vehicle to maintain and promote Polish independent thought and nationalism. The sisters were "flying" at their own peril, but they were willing to assume the risk because they felt just as deeply about receiving higher education and promoting Polish freedom as the Russians were against both concepts.

But as exhilarating as attending the "Flying" University was, with its secret gatherings and earnest lessons taught by selfless, patriotic instructors, the simple truth was that it could not substitute for a formal university education. Four centuries ago, as a man of means Copernicus had taken as his right the opportunity to enroll in the finest universities, first in Poland and then across Europe. Now, for Manya, her sister, and other Polish women, a college degree still could not be obtained in Poland, and they knew they would have to leave the country to do so. Both sisters had harbored secret dreams of getting a degree at Europe's most renowned educational institution, the Sorbonne in Paris. But neither had any idea of what to do about it, that is, until a flash of inspiration hit Manya, as it would many times in the future. She met the problem by applying her inherently logical thinking, somewhat similar to how she would solve a mathematical equation.

Each sister wanted a university education, which cost a significant amount of money that neither had. But both did not require the education at the same time. Why not have one sister support the other with funding over the course of the number of years needed to complete the education? This would make it easier to accomplish, a portion of the support provided each year rather than funding the complete amount in one year. Then, once the first sister had received her education, she could use her degree to get a job and support the other sister to gain an education in the same manner.

Now came a difficult decision, the part of Manya's solution that wasn't calculated by employing a mathematical formula but required personal sacrifice and determination. Manya proposed that Bronya, the older sister, should be the first to receive her education. She would work to support Bronya's university studies, Bronya would receive her degree, and then do the same for Manya. Like a mathematical proof, it balanced perfectly, appealing to both sisters from an intellectual perspective. Tutoring alone could not provide the sums required to support this plan, so Manya informed her sister that she would become a governess for a family in the city, live at home, and send half of what she earned to Bronya.[12] And so, after Bronya's protestations about her sister's sacrifice were overcome, the plan was agreed.

What Manya couldn't have thought to incorporate into her calculations was what she would have to put up with in order to earn her pay. Her first attempt was well-meaning but doomed from the start. A family in Warsaw needed a governess, and she needed a job. A simple transaction. Only when flesh-and-blood human beings were substituted for the idealized family Manya hoped she would work for, the plan quickly turned sour. The young governess and the lady of the house saw eye-to-eye on practically nothing, making each encounter a tension-filled confrontation. Each felt she was above the other in stature, the employer simply by virtue of her position, Manya knowing she was of superior intellect. Quickly Manya realized this job was not worth the effort.

Manya ended the engagement and formulated her next move. She would search for a governess job in the country, where she could live free of room and board expenses, allowing her even more money to save in support of her sister's education. If only she had come up with this idea sooner, she could have begun the process earlier and saved more money for her sister as well as avoided pain and aggravation for herself.

As it turned out, this next approach was altogether different. Manya found a position in short order, only weeks after her initial, failed effort. The family that she was to work for was delightful, with parents who greeted her with welcoming arms and children who were a pleasure to teach. She even developed a close relationship with the older daughter, Bronka. But it took a while for Manya to come out of a self-imposed shell, in part created by her residual feelings of inferiority from her last governess experience, in part due to her newness to her environment and lack of established relationships with anyone in the area.

Habitually, Manya closeted herself in her room after a long day's work, reading texts on mathematics, physics, and philosophy, and solving math problems sent by her father to sharpen her skills, acquiring "the habit of independent work" that would be so valuable to her later in life.[13] Manya knew she would need to augment the education she had already received with extra study in order to lay an acceptable groundwork for her plan of education at the Sorbonne, especially in mathematics and the sciences, which were the most important subjects to her. And her insecurity also took time to surmount, but eventually Manya did overcome it. As she did, she fell in love.

Casimir, Bronka's older brother, who had been attending Warsaw University while Manya was settling into her position, came home at the end of the spring semester. Each was instantly captivated by the other, Manya a sensible, pretty, intelligent girl unlike the flighty women Casimir had met at school, he a handsome university student from a well-to-do family. They soon declared their feelings of mutual love, impetuously planning to marry. They both wanted to quickly take a next step, caught up in their bliss. But, when Casimir sought the approval of his parents, he was rudely awakened to the painful realities of class prejudice. Just as many parents critically view prospective mates for their sons, so Manya was seen as someone who would never be good enough for their beloved boy. Practically penniless, almost a servant in their household, she was not the woman they had pictured for him. Young Casimir, in love with Manya though he was, was no match for the stern rebukes he received from his parents on this issue.

Where Manya was strong, determined to overcome this obstacle for a future together, Casimir was weak, more afraid of losing his parents' financial support for his continuing university studies than of losing the woman he professed to love. Better to find this out now rather than put any more effort into this relationship, another difficult yet important lesson in Manya's ongoing education. The family was content to have Manya continue in her former capacity as their governess, but not as their future daughter-in-law. She accepted this approach for the sake of the money she would earn, trying to bury her love for Casimir and move on. Over the course of the next few years, Manya's romantic feelings were to ebb and flow until, after trying to get him to finally make a commitment, Casimir once again backed away and all chance of a future together ended in her mind. She swore she would never marry.

Time passed, and Manya's financial support of Bronya had allowed her sister to move to Paris to study medicine at the Sorbonne. Her employment in the country came to an end and she then became a governess once again in Warsaw, saving her rubles to send to Bronya as well as to help her father. Surprisingly, while back at home she encountered a situation that surely stimulated her mind. One of her cousins ran a small laboratory in Warsaw, arranging for Manya to periodically slip in alone to utilize the equipment for basic scientific investigations. She noted in her autobiography, "I tried out various experiments described in treatises on physics and chemistry . . . at times I would be encouraged by little unhoped-for success, at others I would be in the deepest despair because of accidents and failures resulting from my inexperience . . . this first trial confirmed in me the taste for experimental research in the fields of physics and chemistry."[14] Years later, Eve Curie recalled her mother's reminiscence of her excitement at running these experiments, coming home at night after an evening at the laboratory, "An exultation different from all those she had known kept her from sleep. Her vocation, for so long uncertain, had flashed into life."[15]

Finally, in the fall of 1891, it was Manya's turn to follow her dream and come to Paris to pursue her university education. Her sister Bronya had graduated from the Sorbonne with a degree as a physician. Along the way, she had met and married another Pole at the university, Casimir Dłuski, a brilliant student also to become a doctor. The couple set up a bustling medical practice in Paris, Bronya attending to female patients, Casimir to the men. Manya journeyed from Poland to Paris and settled into the Dłuskis' home, welcomed warmly by Bronya and brother-in-law Casimir.

In 1891, as Manya registered at the Sorbonne for her first course lectures, she signed her name as Marie, only natural as she strove to be accepted in this most Parisienne of institutions. Dating back to its founding as a theological school in 1253, the university had become famous throughout Europe as an unparalleled center of learning. Now Marie had achieved what she thought might never happen, becoming part of this illustrious school, as one of less than two dozen women among over 1,800 students attending the school of science that year.[16]

As the 19th century was coming to a close, the teaching of the science curriculum at the Sorbonne was in excellent hands. Among others, Gabriel Lippmann, future winner of the 1908 Nobel Prize in physics for his work on color photography, was perfecting his theories and directing the university's physics research laboratories while lecturing on experimental physics.[17] The renowned mathematics savant and theoretical physicist Henri Poincaré, who wrote with mastery on a variety of topics while publishing hundreds of papers and over thirty books, lectured on mathematics and mathematical physics to a captivated Marie.[18] Poincaré's close friend and brilliant theoretician Paul Appell taught analytical mechanics with such elegance and certainty that Marie was to smile with joy as she absorbed his lectures and sincerely wondered, "How could anybody find science dry? Was there anything more enthralling than the unchangeable rules which governed the universe, or more marvelous than the human intelligence which could discover them?"[19] This euphoria concerning her educational experience was a constant for Marie, becoming a trademark of her time at the Sorbonne, an exhilaration that accompanied her thrill at being in the midst of a scientific environment on the theoretical and experimental leading edge of learning. The energy she derived from the lectures she attended would help her through the most difficult of periods during a collegiate career that often stretched her physical stamina to the limit.

Marie had already recognized that she needed to bring her working knowledge of scientific and mathematical subjects up to another level in order to have a firm basis on which to build with the materials that were being taught in her current lecture classes. As well, she was embarrassed by her Polish version of somewhat fractured French as she engaged in daily conversation. Both issues could be addressed by her approach of rigorous study in these areas, which she immediately began, in addition to her already heavy course load. She simply needed more time to devote to all of this, time that could be found if she eliminated the two-hour daily commute from her sister's apartment to the university and back home again, both trips that needed to be taken in a costly omnibus carriage. Finding an apartment close to the Sorbonne would alleviate the problem and be less costly in the bargain.

Besides, Marie found that spending time at her sister's, though enjoyable, provided too many distractions from her studies, what with socializing with Bronya, Casimir, and their friends from the Polish expatriate community in the area who often stopped by in the evenings. She chastised herself, repeating in her mind the singular thought that she had come to Paris to learn, and learn she must. As much as she hated to leave the comfort of Bronya's apartment, Marie found lodgings in the Latin Quarter, the age-old haven of starving writers and artists, prostitutes, and penniless students, within a few minutes' walk of the school. She moved in at once.

Marie was to live in a number of locations in the area, each cheaper than the last, until she finally settled in a room that was six floors up in the attic of a home. There she could study in quiet solitude. Never mind that it was a simple room, devoid of any charm or creature comforts that even a picture on the wall or a rug to cover the scarred wooden floor might provide. She had little money from her savings to spend on housing, which left a tiny sum for her meals and coal for heating the room in winter. A few pieces of furniture on which to sleep and eat, a lamp for evening studies, and a small, coal-burning stove were all she required. The few visitors she had were welcomed to a seat on Marie's clothes trunk when they stayed for more than a few minutes. Occasionally, she even fainted from hunger, Bronya and Casimir always miraculously there to rescue her and nurse her back to relative health at their home. But soon enough, she was back in her prison cell of a room, subsisting on morsels of chocolate, a piece of fruit, a bit of tea, and a slice of bread. Or, as daughter Eve described it, "she began again to live on air."[20]

Yet, Marie was to recall these days as some of the best of her life. Living the spartan existence that she had chosen for herself was a minor impediment on the road to realizing her educational goals. As she remembered this period in her writings, "Undistracted by any outside occupation, I was entirely absorbed in the joy of learning and understanding . . . this life, painful from certain points of view, had, for all that, a real charm for me. It gave me a very precious sense of liberty and independence . . . all that I saw and learned that was new delighted me. It was like a new world opened to me, the world of science, which I was at last permitted to know in all liberty."[21]

Brilliant scholarship teamed with indefatigable determination to fuel Marie's educational achievements. A full day of classes was followed by an evening's study in the library. Then, Marie would come back to her tiny room to continue her work. She would take a few bites of food while reading, applying lessons learned to solve mathematical and physics problems, or plow through French grammar and syntax until the small hours of the morning. Only then would she fall asleep, waking early the next day to repeat the cycle. As with many a superior student, perfection was her standard, and she was sorely disappointed if she failed to achieve this goal. After two years of this punishing effort, she still approached her final examinations with trepidation and anxiety. But these fears were unfounded. In the spring of 1893, she obtained her master's degree in physics, ranking first of those being tested. A year later, she was to receive her second degree, a master's in mathematics, this time finishing with a well-earned, but disappointing to her, second place.

Now that she had become a physics master's graduate of the Sorbonne, Marie had completed what she had set out to accomplish. In the process, she had proven to herself what she could do under the most daunting of circumstances, employing her mind and a will of iron to traverse the deepest rivers filled with challenging cross-currents and unsure footing. If not well-off financially, she was nonetheless becoming somewhat comfortable in Paris. Bronya and her family were nearby, she had developed some acquaintances in the Polish expatriate community, and her French was now impeccable. Her original intention had been to return to Poland after she received her degrees so that she might be with her father as well as bring her learning home to educate her own compatriots. Now, she was not so sure.

Conveniently at this juncture, through the influence of a favorite professor, Gabriel Lippmann, Marie received a grant from a French scientific organization to investigate magnetism as it applied to different types of steel.[22] Lippmann, as head of the Sorbonne's physics laboratories, allowed Marie to set up her equipment to perform this work in a small corner of one of the labs. She lamented to a few Polish friends that the space could not adequately house the various instruments she needed to perform her experiments and measurements. Fortunately, one of the friends knew a professor teaching at another institution in Paris, a physicist who himself was an expert in magnetism and who might have enough room to house Marie's experimental paraphernalia. His name was Pierre Curie.

So it was that a meeting was arranged at the home of these Polish friends, man and wife, to which Marie and Pierre Curie were invited to discuss Marie's situation with her research. In her autobiographical notes, Marie comments on the thirty-five-year-old Pierre as having a "grave and gentle expression of his face, as well as a certain abandon in his attitude, suggesting the dreamer absorbed in his reflections."[23] Their daughter Eve describes him as "having a very individual charm, made up of gravity and careless grace."[24] Both sensed a man comfortable in his own skin, serene in an intelligence that dared to search for answers, unrestricted by convention. And Marie was no stranger to the power of dreams, having so recently achieved her own in earning an advanced degree at one of the most prestigious universities in Europe through her brilliance and tenacity.

They passed a pleasant evening, Pierre unfortunately unable to help Marie with laboratory space since he explained that he actually had none of his own. But, he could offer a wealth of practical advice on the properties of magnetism, and they proceeded to discuss this subject as well as other scientific topics of interest to both, physicist to physicist. The conversation then moved on as the three Poles exchanged information concerning their homeland, at times drifting into their native Polish as they spoke. As he listened to their conversation, even in an unfamiliar language, Pierre was transfixed by Marie's gentle yet direct expression. How marvelous, Curie thought, a scientific woman of the highest intelligence and conviction. Here was someone who understood not only physics and mathematical concepts, but the meaning behind the theories, and who could express herself in multiple languages with unwavering opinion.

As the conversation of the three Polish immigrants proceeded excitedly, Pierre faded into the background, shifted back in his chair, and allowed a vision to superimpose itself on his thoughts. His love of nature, nurtured from an early age, now often resulted in long solitary walks culminating in his return with a handful of light blue chicory, vibrant purple bellflower, and soft yellow marigolds to put in a vase in his parents' home. He suddenly wondered if Marie liked wildflowers.

CHAPTER THREE

Life Partners

The winner of the International Championship tournament on October 4, 1900, was presented with an unusual trophy, a delicate porcelain bowl mounted in chiseled gold. Artistic in nature and valuable in its own right, the award was uncommon for a sporting event. But it somehow seemed appropriate to bestow it on Margaret Abbott, a twenty-two-year-old American participant who had just bested the field of nine other women, two French and the rest Americans, and won the first ever female Olympic golf championship at the Paris Olympics. Margaret had been living in Europe with her mother Mary at the time, and they had come to Paris so that Margaret could take lessons in sculpture and painting. Possibly as a diversion from her studies and the accompanying visits to the seemingly endless art galleries of the city, she decided to enter the golf contest, discovering she could join a few days before it transpired. Margaret had previously played golf for a number of years with consistent success, mainly at country clubs in the Chicago area, where she and her mother had settled after the death of her father. Mary actually entered the event as well, in doing so comprising what would be the only mother/daughter entrants ever to participate in the same event in the same Olympic Games; Margaret's winning nine hole score of a respectable forty-seven beat her mother by eighteen strokes.[1]

These Olympic Games had been organized in conjunction with the International Exposition of 1900 in Paris, a World's Fair held from mid-April to mid-November that year. The Exposition was a tremendous attraction for the city, with close to fifty countries and numerous businesses participating and tens of thousands of exhibits displaying cutting edge technology of the day, including a Palace of Electricity and a moving sidewalk to effortlessly transport the masses from one major point to another on the fairgrounds. It all had lured fifty million visitors during that seven-month stretch, breaking all previous records for attendance to similar expositions. In anticipation of the crowds to come, Paris had busied itself years earlier in beginning to construct a huge new underground metro system, the first section of which was almost ready for the start of the event's opening, being commissioned in July.[2] The Eiffel Tower, a great hit at its inauguration at the last World Exposition to be held in Paris in 1889, shone with a bright golden glow, given a stunning makeover for the occasion with a fresh coat of yellow paint.

Athens had hosted the first Olympic Games in 1896, followed by Paris sponsoring these games in 1900. The competitions were poorly planned, spread sporadically over five months in various parts of the country while the World's Fair was in progress. A dearth of spectators at many events testified to the erratic scheduling and lack of publicity for the contests, some of which, like the women's golf competition, were not even billed as part of the Olympic Games but rather as vaguely framed international contests.

But the Parisian organizing committee did make one monumental decision for these games. They would allow women to participate on a very limited basis, opposing the views of the president of the overarching International Olympic Committee, Baron Pierre de Coubertin. Twenty-two women competed in the events out of 997 participants, representing a minute fraction at just over two percent.[3] Besides golf, the only other all-women's event was lawn tennis, both sports being deemed feminine enough not to cause too much of an issue with the general public. Women were even allowed to participate alongside men in a few of the other, more genteel sports: yachting, equestrian, and croquet. The first woman to win an Olympic event at these games was an intrepid Swiss female sailor, joining with her husband in one of the yachting events, no doubt serving as the faithful crew to his role as captain. The first individual women's champion was from England, tennis singles player Charlotte Cooper. And Margaret Abbott had won the women's individual golf match, though it was so poorly publicized as an Olympic event that even when she died fifty-five years later, she did not know that she had participated in, and won, an Olympic competition.[4]

Pierre de Coubertin was not pleased with these developments concerning women's participation in the Olympics. De Coubertin was an aristocratic Frenchman and the founder of the modern Olympic Games, a vision he proposed during a convention of athletic associations from numerous countries at the Sorbonne in June of 1894, shortly before Marie Skłodowska earned her degree in mathematics at that institution. The other nations rallied around his plan to proceed with an international competition to resurrect the glory of the ancient Greek Olympics in modern times. This was part of a growing global movement toward the end of the 19th century to support athleticism and sport as a symbol of stamina, team spirit, and virility for young men. However, de Coubertin's notion went further, stating that, just as in the original Olympics, there was no place for women in these games, and much of society agreed. They were deemed too biologically and temperamentally fragile for strenuous athletic competition, and were generally encouraged to refrain from physical exertion and to be modest in dress and comportment in public.[5]

Women in other pursuits during the latter half of the 1800s were beginning to force cracks in the barriers that had been constructed by their male counterparts, who had consigned them to a domestic role versus one in the public sphere. The American Revolution had been fought along gendered lines of the Declaration of Independence, stating that "All men are created equal," while the French Declaration of Revolution, thirteen years later, was titled "The Rights of Man and the Citizen," both documents authored without any intent to include women in their enlightened views of civilization.[6] Western culture was just not ready to take the enormous leap of extending what Thomas Jefferson described as the inalienable rights of "life, liberty, and the pursuit of happiness" as they might fully apply to women.

What ensued was a two-pronged effort to justify this gender discrimination, expressed with such overwhelming male authority that it was difficult to overcome. Women's roles were explained to be naturally domestic, while a man's place was public and political, supported by male scientific and medical experts who noted that the biological differences between the sexes dictated these distinctly separate roles.[7] In 1871, Charles Darwin had even commented on what he perceived as obvious gender differences related to intellect in his book *The Descent of Man*. He stated flatly, "[Man] has a more inventive genius . . . The chief distinction in the intellectual powers of the two sexes is [shown] by man's attaining to a higher eminence, in whatever he takes up, than woman can attain—whether requiring deep thought, reason, or imagination . . . the average standard of mental power in man must be above that of woman."[8] Obviously, he had yet to meet anyone like Marie Skłodowska. Even as women were assuming new positions in society and nearing the attainment of suffrage across much of the western world in the early 1900s, an unlikely defender of women's domesticity was a towering pioneer of American investigative journalism, Ida Tarbell. She argued consistently against women's right to vote, maintaining that participation in the public sphere of politics was against the very nature of a woman's domestic mission, in her view wrongly shifting women's focus from their primary responsibility of rearing and stewarding the domestic family unit.[9]

Yet, in the face of this supposed evidence of male superiority and resulting role assignment, women were slowly becoming accepted during the 1800s in areas beyond domesticity. Teaching and nursing were public professions to which women gravitated, being viewed by a broad swath of the population as jobs that benefited by a more "feminine" touch, and so representing possible exceptions to the rule of domestic servitude to which most women were relegated. This began to extend to social work and, for women of independent means, philanthropy, again because of the "caring" nature of these roles, drawing on what were considered feminine characteristics. At the same time, female authors, artists, and performers were more increasingly becoming known to the public as legitimate contributors to national cultures.

As men became more interested in higher education, especially university schooling, so too did women begin to clamor for these opportunities, and with them the possibility of using their hard-won degrees to move into the professional worlds of medicine, science, and industry. In Europe, the trend was a bit slower to materialize than in America, where women-only colleges allowed for higher educational opportunities while still treating the two sexes separately. The Sorbonne in Paris was the second university in all of Europe, after Switzerland's University of Zurich in 1865, to allow female students in a coeducational framework, awarding the first degree to a woman in 1870. By 1890, 210 women were students at the Sorbonne, 111 of them from countries outside of France.[10] The next year, Marie Skłodowska was among the new women enrollees, continuing the trickle of female participants into the mathematics and scientific disciplines.

Simultaneously, a few notable women were developing reputations as outstanding scientific minds. One of the first, and most famous, was Russian mathematician Sofia Kovalevskaya, who was tutored as a woman of privilege from minor Russian nobility and became the first woman university professor in Europe, attaining that position in 1889 at the University of Stockholm. Recognized as an exceptional mathematician by the scientific academies of Russia, France, and Sweden, she aroused considerable controversy by her appointment as a full professor, but continued to win international honors with her ingenious insights into complex mathematics, while at the same time even authoring numerous creations including an autobiographical novel.[11] Sofia was a strong proponent of women's rights, especially related to attaining higher education. She entered an arranged marriage with a Russian man who initially was supportive of her efforts, but the couple grew more distant as time passed, each consumed in following their own professional callings. During their time as husband and wife many paths to potential university professorships were closed to her as a married woman, most European nations subsuming a wife's legal rights under those of her husband. However, upon her husband's death, the widowed Kovalevskaya was now viewed as a respectable candidate to become a professor at a number of European universities. In 1883, four months after her husband passed away, she secured a temporary teaching position at the University of Stockholm, gaining permanence to the post six years later in 1889.[12] She died only two years afterward from a bout of pneumonia, by then internationally recognized as a groundbreaking mathematical physicist.

In England, physicist, mathematician, and engineer Hertha Ayrton was also a champion of women's rights and a Curie contemporary who would eventually meet and befriend Marie. Although she came from modest means, she was able to gain admission to the first English all-women's college, Girton of Cambridge. Hertha married a Fellow of the Royal Society of London, Professor of Engineering William Ayrton, and studied electrical engineering under his guidance. She proceeded to make a number of practical discoveries in this discipline, including methods to more efficiently use the electric arc lighting that was commonly employed in street lamps in many parts of England. At the same time, Hertha began to study water ripple effects on underlying beach sands, the principles of which she employed during World War I to devise a mechanical fan that could adapt these wave effects to air currents to help remove poison gas from battle trenches and later for improved ventilation in mine shafts and sewers.[13] After being nominated in 1902 to become a fellow of the Royal Society in her own right, Hertha collided with the same obstacle being thrown in the path of every woman attempting to become a member of an all-male national scientific association. She was denied this honor simply because she was a married woman, and as such not a "person" in the eyes of British common law, a convention that did not change until 1929 in England.[14] It took until 1943 for a woman to again be nominated, with admission of women into the Royal Society finally taking place in 1945.

What was evident to those who more closely examined the situation was that women were able to achieve to their fullest potential in the scientific disciplines if they were married to enlightened men who encouraged and often financed their spouses' investigative efforts. Such was the case for Ayrton, whose husband had been supportive of Hertha's independent endeavors, pleased with her interest in electrical theory and practical application. Certainly this would eventually be so for the wholehearted support given Marie Curie by her husband Pierre. It was not quite so simple for Kovalevskaya, whose mate quickly reverted from encouraging Sofia's educational aspirations to desiring a more domestic focus for her. It was only after his death that her bonds to a traditional role were completely broken. Sofia as a widow became a woman who had legal as well as personal jurisdiction over her life, accepted in much of society as a "person" in her own right.

The importance of having this control was especially vexing for married women, who had virtually none of this power through well into the 20th century. Laws across Europe and, indeed, throughout most of the world, gave all rights of any earnings and property that a married woman may have had to her husband. In France, the Curies were subject to the Civil Code of the day. This Code clearly stated that married women, children, and the insane were all classified as incapable of "managing their own affairs," including owning property or dealing with any pay they may have received for their work, this section of the Code remaining as law until 1938.[15] Effectively, married women were only allowed financial rights to the extent that their husbands saw fit to bestow any upon them. In this environment, a supportive spouse was a cherished exception rather than the rule for women at the turn of the century.

Somewhat buried under the myriad activities during the summer of 1900 in Paris, while women were breaking through the participation barrier in Olympic sports and other aspects of culture, Marie Curie was involved in a gathering that, in its own way, was almost as novel. The First International Congress of Physics was brought together by the French Physical Society, one of over 120 various organizational conferences across a wide range of professional pursuits that were scheduled in Paris that year to take advantage of the drawing card of the World's Fair. In fact, Hertha Ayrton attended the International Electrical Congress in Paris that year as well, giving a paper on her electric arc lighting investigations.

Just prior to the conference that Ayrton attended in late August, the sessions for the physics conference took place during five days early in the month, attracting 836 attendees from across the globe, estimated to comprise well over half the world's degreed physicists.[16] Numerous papers were given to the international audience, each describing the state of discovery in a variety of fields of physics at the turn of the century, from electricity and magnetism to cathode rays and the atom. One after another, over eighty male scientists gave informative, often weighty presentations on their specific areas of expertise. Only two women were part of the crowds at the convention that week, Marie Curie and Isabelle Stone, the first American woman to obtain a physics doctorate degree.[17]

Marie had come along with husband Pierre while he gave a keynote presentation on the astounding phenomenon of radioactivity that they had discovered two years prior as they investigated the mysterious source of rays emanating from uranium. This was a curiosity previously observed by Henri Becquerel in 1896 as he was doing experiments on the newly discovered Röntgen rays, or X-rays as Wilhelm Röntgen called them. Pierre presented the paper, describing in some detail the laborious efforts he and his wife had taken to determine that these rays were not the province of uranium alone but were also properties of other heavy elements that they had subsequently discovered, polonium and radium. Although he stressed the team approach taken by the couple, the attendees couldn't help but assume that it was Pierre who provided the intellectual spark, with his wife as able assistant. Marie heard a murmur of excitement ripple through the crowded conference room as Pierre told of their current efforts to isolate pure samples of the radioactive materials. She smiled as she observed her husband humbly yet clearly conveying the details of the exhaustive investigations they had conducted together that had led to this point, and she reflected on all that had happened since first meeting Pierre that fateful evening at their mutual friends' home six years before.

As Marie studied furiously to prepare for her examination in mathematics in the spring of 1894, she also worked on her project analyzing magnetic properties of different types of steel. Although she had found that Pierre could not provide her with laboratory space for her testing, they spoke often about her work. His continued interest in both this attractive woman as well as her project led to his coming to visit Marie in her dreary room, where he could see just how dedicated she was to her studies and testing as she made him a cup of tea and brightly conversed about technical issues concerning her experiments, completely oblivious to her sparse surroundings. The more he saw of her, the driven physicist, pure in her efforts to achieve her goals, the more he saw of the same obsessive devotion to science that he, himself, possessed, two people completely dedicated to their investigations and the wonders they might unlock for themselves as well as humanity. Magnetism had certainly drawn them together in more ways than one.

But just as Marie's long-ago experience with Casimir back in Poland had resulted in her forswearing a future that included marriage, so Pierre had similarly experienced rejection in a previous relationship that had cut so deeply that he felt he had permanently moved on from love. That is, until now. As he continued to visit the lonely garret, where a cup of tea and scintillating scientific conversation awaited, Pierre felt a growing attraction. As they learned more about each other, they found that they shared an eerily similar family history, steeped in scholarship and parental nurturing, which brought them even closer together. It became apparent that many of their personal characteristics complemented each other as well. Where she was intense, he was capricious, she more energetic to his natural introversion. It wasn't long until their conversation turned to the future and, at least expressed by Pierre, how two people sharing such an intense interest in exploring the scientific world might devise a means to share it together.

Marie was hesitant to accept the proposals of marriage offered sincerely by Pierre. Undeterred, over the next few months of summer, as Marie left Paris to visit her father and family in Poland and contemplate what the future might hold for the two of them, Pierre devised alternative paths they might take, as surprising as they were logical. One was to offer to marry Marie, give up his position in Paris and move to Poland where she could be close to her father and teach those searching for higher learning in Warsaw. He would continue his scientific research and find work somehow, in effect taking the culturally unheard-of step of becoming a male "trailing spouse" in their working lives. This avenue would satisfy the inclination Marie still harbored to bring her newfound university knowledge back to her home country in an effort to elevate the education of her people in scientific principles to better prepare them in their struggle to be a free nation. Another option, even more audacious, was to forgo marriage altogether and live in a shared apartment, separated by a divider of some sort for privacy but joined by their mutual pursuit of science.[18] Pierre's most fervent thought was summed up eloquently in a line from one of his many letters to Marie that summer, the dreamer at his most impassioned, "It would be a fine thing, just the same, in which I hardly dare believe, to pass our lives near each other, hypnotized by our dreams: *your* patriotic dream, *our* humanitarian dream, and *our* scientific dream."[19]

A year later, in July of 1895, Marie and Pierre married. By then, she had been convinced by his constant entreaties, pragmatic and loving at the same time. Before they wed, Marie encouraged him to belatedly present his doctoral thesis to an examining board at the Sorbonne, a treatise on magnetism that encompassed his painstaking work on how temperature affects magnetic properties of various magnetic materials. Pierre's work was recognized by those on the board as being a project of superior design and detailed investigation that exhibited novel results. It introduced what was later to become known as the "Curie temperature" into the scientific lexicon, a temperature at which many magnetic materials lose their permanent magnetic properties and above which they become only able to re-magnetize in the presence of another magnetic field. Knowledge of this property was to eventually be the basis for the invention of items as varied as specialized temperature controls and magnetic-optical data storage products for computers.

By the time of his presentation, Pierre was already a recognized expert in magnetism. Previously, he had made a number of other scientific discoveries in his life, in his early twenties working with his older brother Jacques, also a physicist as well as mineralogist, on experimentation with crystals and their electrical properties. Of particular interest to the siblings were the intriguing results they encountered from putting certain crystals under physical pressure. They observed that when they compressed some crystals, especially quartz, this produced a weak electrical current that could be measured. Conversely, when an electric current was applied to perfectly formed quartz crystals, they proceeded to become irregular in shape. They had discovered a phenomenon that they named piezoelectricity, the term *piezo* deriving from the Greek for "to squeeze."

Utilizing these findings, the brothers constructed an electrometer, a device that could measure the exceedingly small amount of electrical current generated by the crystals being compressed with weight placed on them. Using quartz crystals, they made an instrument they called a piezoelectric quartz balance. This piece of equipment would become vital to Marie as she sought to understand radioactivity fifteen years later. The current generated from the pressurized quartz crystals in the device could be balanced with an exact amount of electricity supplied by another source, thus providing an accurate measure of the electricity generated by the compressed crystals. Or, conversely, a measurable amount of electricity generated by the compressed crystals could balance the amount of electricity made by another source.

All this was unique and exciting to the Curie brothers, as it was to the scientific community at the time. Measuring tiny changes in electrical current generation was made possible by these devices, and the Curie brothers proceeded to design and manufacture piezoelectric quartz balances, as well as other pieces of measuring equipment, which they patented and sold commercially.[20] Licensing this patent work for others to use to manufacture and sell the instruments actually generated a stream of annual income of about fifteen hundred francs for Pierre, a tidy sum for the effort.[21] But interest waned in further work on piezoelectricity itself as the phenomenon had no obvious, larger commercial applications.

Eventually, this lack of focus on piezoelectricity would dramatically change. In the years to come, Marie and Pierre would employ the piezoelectric quartz balance in their historic discovery of radioactivity to help measure minute quantities of electrical current produced when radioactive substances caused electricity to be generated from electrons in the surrounding air by its ionization. This crystalline piezoelectric effect was to find even more uses as the basis of pioneering innovations from sonar technology first employed in World War I to the use of quartz crystals in clocks and wristwatches still sold today.[22]

However, even with this scientific background of impressive accomplishments, Pierre never sought recognition or reward beyond the self-satisfaction of his successful scientific revelations. He was a loner who rarely extended himself beyond his strong family relations with parents and brother, failing to cultivate professional networks, letting his accomplishments speak for themselves. He had been homeschooled until the age of sixteen, when he finally ventured beyond the family circle and attended the Sorbonne to receive a degree in physics. He possessed a superior intelligence, but one that required intense focus on a particular challenge in order to keep his mind from wandering. And once he had accomplished his goal, he had no great need to impress others with his results or seek their approval. His was a life devoted to science, to the investigation of the unknown with the distinct reward of achievement derived from the often-intriguing findings that awaited at the end of a journey of exploration. He enjoyed sharing what he had learned through teaching it to eager students, albeit at a minor school of higher education in Paris, the School of Industrial Physics and Chemistry (EPCI). His relatively introverted nature had kept him from developing significant connections with major influencers who could help him secure a professorship at the more well-known institutions like the Sorbonne. But now he had found someone with whom he could share his explorations, someone who was of a similarly brilliant mind and attitude toward scientific investigation as well as the world. Their daughter Eve saw the union as only a loving child could explain it: "Two hearts beat together, two bodies were united, and two minds of genius learned to think together."[23]

The two settled into married life wholeheartedly. Marie tackled domesticity just as she would take on any new venture, with logic and her patented determination. In many ways, it was like a whole new course load of subjects in school. Learning to cook, an art she had failed to spend any time on in her young life so far, was like taking a chemistry course, enhanced by the pleasure, or peril, of eating one's experimental results. Living on a budget required mathematical accounting skills, balancing income and expenses, if possible. If not, learning how to economize to make ends meet was the only possible alternative, at least in her mind.

Pierre continued to teach and do research, joined by Marie as she completed her experiments on metals and magnetism. Once she passed her examination for teaching in 1896, both she and Pierre often spent their time together discussing scientific topics, reading, and preparing lessons for their students. "My husband and I were so closely united by our affection and our common work that we passed nearly all of our time together," she fondly recalled.[24] Every so often, when time permitted, they loved taking bicycle rides into the countryside as well as longer vacation excursions with bikes in tow. They had received a wedding gift of money enough to purchase two of the delightful contraptions for their honeymoon, when they took their first bicycle adventure together. With both enjoying the outdoors and the excitement they found in exploring nature, it was a perfect activity for shaking off the grinding hours of mental focus on books, theory, and scientific experimentation.

By 1897, Marie had finished her project on metals and magnetism. But the delicate balance of life at the Curies' was to be complicated significantly by two new developments. Marie gave birth to their first daughter, Irène, followed within a few short weeks by Pierre's mother dying and his father coming to live with them. They welcomed the former, and had unfortunately expected the latter. These life cycle events added new components to their lives, which could have made things untenable had not Pierre's father stepped in to become a surrogate parent to the newborn girl. This allowed both Marie and Pierre to continue their work with minor schedule adjustments. Pierre's father, a physician, took great pride in his relationship with Irène, forging a lifelong bond and encouraging her interest in science that was to last a lifetime. His arrival in the household, sad as it was with the passing of his wife, could not have been better for Marie in the long run, allowing her enough breathing room in her schedule to contemplate her next educational endeavor: attaining a doctorate in physics.

As Marie and Pierre discussed this next step in scholarship, uppermost in their minds was the topic on which to base Marie's doctoral dissertation. Scientific investigation was most compelling when delving into the unknown, exploring an area that had received scant focus to date so that resulting experimentation might yield truly novel results. Still, few deviated from the well-trodden, safer path of working to expand on new discoveries of others, where efforts still could produce nuances yet to be unearthed.

Only a few years before, in 1895, Bavarian physicist Wilhelm Röntgen had astonished the world with the discovery of X-rays. To fascinating effect, he exhibited for all to see the product of a photographic plate that showed the skeletal outline of the bones in his wife's hand, complete with a prominent ring, after he shot X-rays for a number of minutes at her hand on the plate. The rays, a stream of electromagnetic radiation emitted from a cathode ray tube, were absorbed to differing degrees by the hand versus the flesh, the calcium in the skeletal bones absorbing it almost completely while the surrounding soft tissue of the hand let it mostly pass through, generating a more shadowy image. The photos themselves were termed "shadowgrams" and the rays producing them designated with the term "X" to signify the unknown. The concept of rays that could penetrate skin but clearly show more solid parts of the body encased within was astounding, leading to ideas of their use both practical and ludicrous. Broken bones could plainly be identified and mended with more precision and efficiency using X-rays. At the same time, speculation about use of the rays for voyeuristically penetrating clothing to reveal the details underneath scandalized polite Victorian society.

Many scientists in Europe were quick to begin researching the X-ray phenomenon. Cambridge's Cavendish Laboratory physicist J. J. Thomson and his graduate student, New Zealander Ernest Rutherford, were some of the first. Thomson conducted his own cathode ray experiments in 1897, resulting in the discovery of previously unknown negatively charged particles that the rays produced that he termed electrons. A horde of researchers in France began to investigate the X-rays further, causing over 65 percent of the papers presented at the French Academy of Sciences in the next few years to be focused on the subject.[25] A herd mentality had taken hold, many scientists anxiously joining the growing stampede to study the famous and mysterious X-rays in the hopes of determining more about their origins and nature before anyone else.

Although Marie was tempted to follow this avenue of exploration for her thesis, she ultimately decided on an alternative that had held some promise in the initial rush to learn more about the nature of the X-rays but had since been discarded. Henri Becquerel, a French physicist from an esteemed scientific lineage, had found that compounds of uranium appeared to emit rays curiously similar to X-rays in that they could penetrate matter, but they produced somewhat hazier images on photographic plates. Although this finding was indeed novel, and he immediately reported it to the Academy of Sciences in Paris in several papers over the next few years, it generated only passing interest with a few other physicists and Becquerel went on to pursue other avenues of scientific investigation. Neither he nor anyone he had presented his finding to at the Academy realized that Becquerel had discovered the phenomenon of radioactivity with his investigation of what often were later termed "Becquerel rays."[26]

Marie stated in her autobiographical notes that Becquerel's work, though it appeared to have been adequately explored, was a subject that required further investigation. "The next step was to ask whence came this energy, of minute quantity, it is true, but constantly given off by uranium compounds under the form of radiations."[27] Like X-rays, the radiation emitted by uranium "ionized" the surrounding air, producing a tiny electrical current by causing electrons in the air molecules to be released. Marie proposed to quantify the amount of electricity being generated, employing very precise but difficult-to-use measuring devices, especially the instrument that her husband had invented more than a decade before, the piezoelectric quartz balance. Once they agreed on the general approach to her research, through Pierre's school they obtained a small space for Marie to start her investigations. Though it was crowded, unbeknown to the Curies, this was where scientific history would be made.

By mid-December 1897, Marie had begun her efforts, learning how to manipulate her delicate measuring instruments to obtain accurate results. They required a high degree of physical coordination in using the devices as well as intense concentration in the measuring of such minute electrical currents generated by uranium. Marie proved quite adept in the former, and the latter characteristic was inherent to her very being, so she was certainly well equipped for the task before her.

Pierre instructed Marie in the use of these instruments, especially the piezoelectric quartz balance for which he was the inventor and expert. Soon, based on her growing proficiency with the equipment, she was testing uranium as well as a full range of other known elements. And, as she became more comfortable with handling the devices, she would pick up the pace and test a number of items in rapid succession. This was her version of today's laboratory practice of high throughput screening, the repetitive testing of many materials at once to speed up the process. Marie examined each sample quickly and efficiently, her deft handling of the various pieces of complex equipment combining with her hallmark intense concentration to yield productivity similar to a one person assembly line.

In her testing, which broadened to include more than just pure elements themselves, Marie now examined a few ore samples in which uranium was already an identified component. Some results were perplexing. One item in particular, a mineral ore called pitchblende, provided an unusual outcome. Pitchblende was mined in central Europe as a known source of uranium for ceramic glazes; testing it showed a generation of electrical current that far exceeded that caused by the emissions of pure uranium.[28] Another uranium-containing mineral, chalcolite, behaved in a similar manner. A third mineral ore, aeschynite, which did not contain uranium but had elemental thorium in it, gave off a current that was also stronger than uranium.

Marie's investigations were producing results that were as mystifying as they were intriguing. Unknown substances within the mineral ores pitchblende and chalcolite were producing rays stronger than the uranium rays Becquerel had discovered, while aeschynite clearly showed its thorium content was responsible for strong emissions as well.

Marie now focused on the pitchblende ore. Removing the uranium from the ore still left material that emitted rays much more powerful than pure uranium.[29] Since only uranium and thorium had registered any electrical current when tested, and the pitchblende being investigated, even with the uranium removed, gave off much stronger rays than either of these materials themselves, only one inference could be drawn. The conclusion was clear to Marie, and it was certainly an exciting one: a completely new chemical substance must be causing these emissions.[30]

Now the scientific chase was on and Marie was more anxious than ever to find out where it led. "I had a passionate desire to verify [my] hypothesis as rapidly as possible."[31] Once she explained it all to Pierre, he wasted no time in immediately joining her to help track down the origin of this fascinating phenomenon. He teamed with his wife beginning in March 1898 to delve deeper into the situation. The excitement generated by Marie's intriguing test results was to power them further in their investigations for months, even years, to come. During that time, Marie developed the distinctive nomenclature that described their work, soon to become ubiquitous in the scientific community and eventually with the general public. She coined the unique term "radioactivity" to characterize the rays powered by the substances she had positively tested for this activity, the materials themselves being designated by her as "radioactive."

Marie quickly assembled her findings in a paper that Gabriel Lippmann presented to the French Academy of Sciences a month later, in April 1898. Lippmann was a staunch supporter of Marie and was pleased to offer her findings for consideration by the Academy.

He admired her thoughtful intelligence, attention to detail, and strict work ethic, all of which would eventually lead him to nominate her for fellowship in the Academy in 1911. Presentation of her discoveries was a service he could render simply because he was a Fellow in the Academy, and as such was able to give the paper in Marie's place. Typical of the times, since no woman was allowed to be a Fellow in this all-male association, this was the only route available for her to present these extraordinary findings. In her paper, Marie stated that: "All compounds of uranium were active, that the activity was proportional to the amount of uranium contained, and that the compounds of another element, thorium, were 'very active.' In addition, 'Two minerals of uranium: pitchblende . . . and chalcolite . . . are much more active than uranium itself. This is very remarkable and leads one to believe that these minerals may contain an element much more active than uranium.' In short, Curie had identified a second known element, thorium, as radioactive and also predicted the existence of a new radioactive element."[32]

Marie had arrived at these conclusions from her independent work that took place before Pierre joined her in the further scientific exploration they would make together and for which they would become world-renowned. Later in the year, as the pair began to refine pitchblende ore into separate fractions, Marie developed the astounding theory that science takes for granted today, but which was so earth-shattering when first postulated. She simply stated that radioactivity was a property inherent to the singular element that was generating these rays rather than ascribed to chemical reactions of the material being investigated.

By the end of the year, two new elements were theoretically identified by the Curies due to the intense radioactivity they generated, polonium, then radium. Laboriously crushing samples of pitchblende, then chemically subjecting them to a variety of separation techniques to find the unknown generators of radioactivity greater than the uranium contained in their sample, both Marie and Pierre worked to painstakingly isolate the sources. The first, polonium, which they named in honor of Marie's native Poland, was bound up in tiny amounts within a fraction of the pitchblende that was composed of compounds of the element bismuth; it produced rays hundreds of times the strength of pure uranium.[33] The second was designated radium by the Curies. This was based on the very powerful rays generated by the element, the name radium from radius, "ray" in Latin. The radium was found in traces of a portion of the mineral that was made up of elemental barium and its compounds, emitting rays much more powerful than those of polonium.[34]

By 1899, the scientific community was spellbound with their papers on radioactivity and the elements that generated it. The Curies' work was presented to the French Academy via Lippmann and another advocate, the original explorer of the mystery rays generated by uranium, Henri Becquerel himself, another Fellow of the Academy. Both presenters gave much-needed gravitas to the theories being postulated by the relative newcomer, a newly degreed, immigrant Polish scientist named Marie Skłodowska Curie and her husband, the respected but reclusive physicist Pierre Curie.

Now an even more difficult task lay before them. In order to chemically prove that these elements were not just theoretical but real, the Curies needed to physically isolate large enough samples of the actual elements themselves to demonstrate conclusively that they existed in tangible form, not just show that bismuth or barium compounds from pitchblende contained hypothesized traces of radioactive elements. To accomplish this, they needed to work on a larger scale, often processing close to fifty pounds of pitchblende ore at a time, thousands of pounds in a year, in order to have a realistic chance to harvest enough radioactive samples to physically verify the existence of these elements. A laboratory with more room for their processing and testing equipment was now required to adequately proceed with this work.

They moved to a somewhat larger, weather-beaten structure close by. For the next three years, their cozily cramped scientific hovel became as much of a home to the Curies as their current living quarters. In a recollection similar to her reminiscences of her student days living in a garret, Marie fondly recounted much of the time spent in the shabby laboratory as some of the most wonderful moments of her life. For her, the surroundings in which she plied her craft, whether in study or in scientific investigation, were never really at issue. It was the achievements that were accomplished in that space that were of significance. Now with Pierre close by, the pair sharing in the excitement of their exploration of the unknown, she was truly content. When one became frustrated, the other was there to offer an understanding word and a compassionate smile. Marie continued her testing and employed various separation techniques required to isolate traces of radioactive material, while Pierre contemplated the next steps in the exploration of radioactivity, diagramming future experiments and puzzling out equations he would scribble down on a nearby chalkboard.

The two had decided to split their workload. Marie had eagerly taken on the chemical separation of the radioactive elements from pitchblende, applying her stalwart resolve, sure in her quick-paced execution of the testing procedures and chemical separation capabilities. Pierre concentrated on the physics of the perplexing rays being emitted. Marie found herself physically straining with the continual grinding of large quantities of pitchblende ore for further processing. This material would then be combined with various quantities of liquid chemicals and heated to separate the radioactive portions from the rest. Next, Marie would employ her practiced laboratory techniques in distillation of the precipitated remains to arrive at minuscule quantities of crystallized radioactive residues. The goal was to build a sufficient quantity of the ultra-pure radioactive salt for analysis and verification that they had indeed discovered a new element.

As she described it, "Sometimes I had to spend a whole day mixing a boiling mass with a heavy iron rod nearly as large as myself. I would be broken with fatigue at the day's end."[35] She would gather enough of the residues to repeat the process, continually re-refining the material. Her hope was to purify the product to a level that could be designated as elemental through its spectral analysis, a means of determining elemental composition by heating the element until it became a gas, then passing the light emitted by the gas through a prism to show a spectral pattern unique to the element.[36]

As they agreed, Pierre was more focused on understanding the nature of the rays being emitted as well as their potential uses. This was a natural division of responsibilities in order to increase the efficiency of their overall workload. It was never viewed by either as an assignment of more cerebral physics investigations to Pierre and more repetitive, chemistry tasks to Marie. But, in reality, both knew that they could not have proceeded in any other fashion. For the practicality was that at this point in their investigations, as they tried to accelerate the overall pace of the effort, in all likelihood if Pierre had taken charge of the chemistry side of the affair, his splendid but often wandering mind might quickly have impeded the effort rather than enhanced it.

As their crude refining process continued, the couple was amazed to see that the tiny quantities they were making produced radiation hundreds of times greater than pure uranium if they were isolating polonium, an order of magnitude greater if they were precipitating radium residues. The presentation Pierre was reviewing with the captivated audience of fellow physicists at the International Congress of Physics in 1900 had now brought the attendees up to speed on all of these developments, as well as covered highlights of the investigations of a growing throng of fellow scientists who had been intrigued by the published efforts of the Curies on radioactivity. In fact, as Susan Quinn noted in her detailed biography of Marie Curie, "It was known that the [radioactive] elements could 'induce' radioactivity in other substances—and indeed had turned the Curies' lab radioactive."[37] This was an unexpected property, one that held as yet unknown signs of danger for the couple, who spent long hours exposed to the radioactivity in their poorly ventilated shack.

Listening to her husband recount their efforts to the crowd of rapt listeners in the audience who were hanging on his every word, Marie was proud of the progress she and her husband had made thus far. Scaly, hardened fingertips from one of her hands, the ends darkened by radiation burns from constant unprotected handling of radioactive materials, drummed softly on the sleeve of her habitual black work dress as she listened to Pierre's explanations of their project's details. Twenty minutes into his speech she was already anxious to return to their laboratory, knowing their work was not nearly close to completion.

It was to take another two years of exhausting labor for her to be able to physically obtain enough of a sample, a tenth of a gram of extremely pure radium salt, to prove its physical existence with a spectrograph unique to the element. There were moments when her husband felt they might be wasting their time, beginning to doubt that they would ever be able to make enough radium to prove its physical existence. Marie, on the other hand, was certain that they were making enough progress to continue down this hard road. Eventually, her persistence was to deliver a discovery worthy of the effort.

As time passed, the Curies couldn't help but wonder what physical manifestation might be the result of their refining labors. They envisioned that the pure substance would generate a captivating color, radium eventually showing itself as "spontaneously luminous," gleaming as it was stored in glass vials with a "phosphorescent blue."[38] Pierre, the intellectual dreamer, and Marie, the brilliant pragmatist, equally shared a fascination with the natural glow of radioactivity that they finally encountered. They often took to carrying tiny, glassed radium samples in their pockets, periodically showing them to dinner guests or attendees at various receptions, the ethereal radiance of the element eliciting childlike stares of wide-eyed wonder.

Early on in their work, in 1899, Marie herself had actually developed a novel speculation on the very nature of radioactivity, having demonstrated that it was not due to chemical reactions of the radioactive materials with other molecules. Her granddaughter Hélène Langevin-Joliot, a noted nuclear physicist in her own right, quoted Marie in sharing her hypothesis that "the radiation is an emission of matter while at the same time the radioactive substance loses weight."[39] She was the first to propose this meaningful insight, which essentially stated that the radioactive atom was degrading in some fashion. But, she couldn't convince her husband of this hypothesis. It took another three years for Marie's atomic theory to be shown as the probable source of radioactive emissions. This was proposed by the physicist Ernest Rutherford, who had left Cambridge for McGill University in Canada, and his assistant Frederick Soddy, rather than the Curies. In fact, Pierre and Marie continued to search for other explanations before finally accepting Rutherford's theory of transmutation of radioactive materials, the natural breakdown of these heavy elements via radioactive emissions.[40] Even though, around this time, the Curies had identified that radium spontaneously released a huge, unexplained amount of heat on a constant basis, signifying internal machinations of some sort, they were reluctant to look for explanations derived from atomic breakdown rather than external forces causing this phenomenon.

The years of equally shared effort Marie and Pierre poured into the pursuit of investigating the phenomenon of the "Becquerel rays" exemplified the scientific partnership they had always envisioned for themselves. Pierre repeatedly needed to explain this to many in the scientific community, never more so than when an honor as special as the Nobel Prize was proposed to be awarded for the investigations that culminated in the revelation of radioactivity. Pierre was given advance notice by a Nobel Prize committee member that he and Henri Becquerel would be receiving the 1903 prize in physics for their amazing explorative efforts, the award to be given without even a mention of Marie's contributions. Pierre would have none of it, explaining that Marie must rightfully be included in the award. This demand would eventually be met, Pierre and Marie receiving the prize along with Becquerel. But the fact remained that many in science could not begin to accept that Marie Skłodowska Curie was a contributor in any significant part of this endeavor. How could she be, they reasoned with the culturally infallible logic of 1903 Europe—she was a woman.

The discovery of radioactivity, and its accompanying elements of polonium and radium, were to catalyze the lives of the Curies, transforming their humble scientific existence into one of increasing fame, financial reward, and job opportunity. First, in mid-1903, Marie successfully presented her findings at the Sorbonne for her doctoral dissertation. Then, in addition to sharing in the 1903 Nobel Prize in physics with Henri Becquerel, which carried with it seventy thousand francs for the couple, Marie received sixty thousand francs herself as part of the Osiris Prize of 1904. This was the lion's share of a one-hundred-thousand-franc philanthropic gift for scientific achievement that was also awarded to Édouard Branly for his work in telegraphy. International acclaim continued beyond the Nobel award when the Royal Society of London, the most prestigious scientific association in England, presented the pair with their highest honor, the Davy Medal, including a monetary award of one thousand British pounds sterling. By 1904, Pierre was elevated from the position of lecturer in the faculty of sciences at the Sorbonne to full professorship, after two previous attempts to achieve this designation had failed due to lack of enough influential connections within the organization. And in mid-1905, he was inducted as a member of the illustrious French Academy of Sciences, the male-only French scientific association at the pinnacle of the profession. Around this time Pierre's name had even been touted as a potential recipient of the French Legion of Honor, a tremendous distinction. With Marie and himself still doing experimentation in their shabby quarters at EPCI, he characteristically refused the offer, responding, "that I do not in the least feel the need of a decoration, but that I do feel the greatest need for a laboratory."[41]

Although they appreciated their good fortune, Marie and Pierre detested the time the publicity of their discoveries now took away from their continuing research. Pierre could finally give up his teaching position at EPCI to his able protégé Paul Langevin so that he could concentrate on his Sorbonne professor's duties. But society now had an insatiable appetite to hear more about the Curies and their amazing efforts. Scientific consultations and conferences with fellow physicists on the intricacies of radioactivity beckoned. This, coupled with continual journalistic requests for greater insight into the Curies themselves, resulted in drastically reduced worktime. The world marveled at the scientific couple that, as equals, had professionally explored radioactivity and yet maintained domesticity enough to raise a young daughter in a loving home. Photographers could not get enough of taking pictures of the heroic couple, legendary overnight but firmly eschewing the trappings of celebrity whenever possible.

The Curies liberally shared their mounting monetary compensation with family and friends, providing gifts to Marie's sisters Bronya and Helena, as well as Pierre's brother Jacques. They could even afford to hire some laboratory staff for their continuing explorations, helping to relieve the burdens of daily laboratory work as the Curies spent more time trying to accommodate the public's fascination with their work. And, in December 1904, their second daughter, Eve, was born, a welcome addition as well as a distraction from the incessant fame of radioactivity.

This sudden appearance on the world's stage was disconcerting to the Curies. Adding to the moment was the increasing discomfort that they were beginning to physically experience, which we can now so easily understand was caused by their continual exposure to radiation. Radioactivity from a powerful source such as radium was proving to be a strange and at times perplexing property, one that presented a marked dichotomy. When radium's beams were employed in a carefully focused manner, it was demonstrated that its emissions could attack and destroy cancerous growths while doing minimal damage to surrounding healthy tissue. At the same time, it could be extremely toxic to internal organs, bone, and even blood if those handling radioactive materials did not do so in a protected manner. Both Marie and Pierre were disarmingly casual in their handling of radioactive elements, physically processing and manipulating large samples of pitchblende ore and resulting radioactive residues that had been concentrated to develop high-purity samples of radium salts.

At the time, they felt little reason to view radiation as the cause of any major physical problems, beyond their fingertips being desensitized sometimes to the point of painful rawness as they were exposed most intimately to test tubes, vials, and vessels containing radioactive particles. Their energy was often depleted, which they could easily ascribe to the long hours they kept, divided between their dedicated research, diligent teaching, and raising a family. Pierre especially experienced shooting leg and back pains as their investigations wore on, usually diagnosed as rheumatism, becoming so painful that they were the cause of his postponing his 1903 Nobel Prize acceptance address until the spring of

1905.[42] Pierre's colleague Paul Langevin noted this leg pain was so severe as to even make walking difficult for Pierre—in fact, radiation exposure can now be presumed to be the origin of the problem.[43]

On the morning of April 19, 1906, the skies were gloomy with dark clouds that doused Paris pedestrians with intermittent, heavy showers. Pierre was one of thousands of locals going about their business in the city, umbrella in hand as each tried to avoid the rain as well as navigate the bustle of traffic mixing people with horses and automobiles on the busy Parisian streets. He was rushing from one meeting to the next at midday after lunching with a group of fellow scientists. Straight ahead in his path was a slow-moving horse cart, ambling close to the sidewalk at a busy intersection as Pierre looked to find the best option around it to cross the street. He chose to go left, around the back of the cart as it approached the intersection on his right. Perhaps his mind started to drift a bit, picturing the faces of the people he would meet at his next appointment and beginning to think about what the session might hold. The rain suddenly commenced again, and he unfurled his umbrella and picked up the pace as best he could, his legs sending now-accustomed pains shooting down the limbs as he increased his stride to quickly cross the rain-slicked street.

As he came out from behind the cart, a large horse-drawn wagon flashed into his view, barreling toward him, coming down an incline to his right with accelerating speed. The wagon's driver barely had time to see Pierre darting out from behind the cart. The man pulled hard on the horses' reins, hoping to slow the animals and lessen what was certain to be a forceful collision. Pierre tried to come to a sudden halt himself and his legs, not responding as they would have in years gone by, flew out from under him and slipped under the vehicle on the wet road. He tried to grasp one of the horses closest to him but fell behind the spooked animal. Somehow he eluded the horses' pounding hooves and the front set of wagon wheels passed harmlessly in front of Pierre's now-prone body. But the back left wheel would not miss its mark, gruesomely delivering instant death by crushing his skull with thousands of pounds of weight from the cargo of military uniform materials in their cases, neatly packed in the wagon above. Everyone could see it was an accident. The roadway was slippery, the congested intersection a mass of moving people, animals, and vehicles. Pierre had emerged from behind a cart, coming out of nowhere, said the wagon driver. The crowd that quickly gathered included the police, all of whom stared at the crushed skull of the victim. A policeman found identification papers on the body. A hushed murmur swept through the crowd—the dead man was Pierre Curie.

That evening, Marie returned home from a day out with her daughter Irène. She was met at the door by Paul Appell, her long-ago instructor at the Sorbonne who had become good friends with the Curies, his eyes red-rimmed and puffy with tears. She glimpsed her father-in-law and another family friend, Jean Perrin, standing in the shadows as Appell quietly told her of the tragedy. She was too stunned to react with the grief that would overtake her in the weeks ahead. Her wonderful Pierre was never to be by her side again. Over the next few days, Marie robotically went about the business of burying her husband, devastated by events out of her control. The heaviness and despair that had swallowed Marie whole so many years before, when in quick succession her sister and then her mother had died, was once again upon her, enveloping her in a sadness so intense that she felt she was drowning.

Of course, as always, Marie would soldier on. The French government graciously offered to take care of her and family with a lifelong pension, but she would stoically decline it, claiming she was an able-bodied woman who could take care of herself and her children. The Sorbonne, after quick deliberation, decided that Pierre Curie's seat on the Faculty of Sciences could most logically be filled by Marie, his assistant at the university. Unlike the pension, she did not rebuff this offer, becoming the first woman professor in the history of the institution.

The couple had experienced an unparalleled scientific journey together. Fellow radio-physicist Frederick Soddy was to state that "Pierre Curie's greatest discovery was Marie Skłodowska. Her greatest discovery was . . . radioactivity."[44] Although accurate, this sentiment covered only part of the story. In true partnership, it could just as easily have been said that one of Marie Skłodowska's greatest discoveries included Pierre Curie. For each was instrumental in bringing out the best in the other, Marie gently but insistently pushing Pierre to realize the potential that lay beneath the brilliant, yet distracted, surface, Pierre helping Marie to temper the blazing desire to achieve with the contentment that scientific exploration in its own right could offer.

Now Pierre was no longer there to provide the perfect counterbalance to Marie's very nature. In the months ahead she would agonize over the loss of his reassuring presence, a person who knew her as no one else ever had. Shortly after his death, Marie began to keep what would later be termed a "mourning journal" containing a handful of personal reflections concerning Pierre's death and her struggle to continue without him. In it, a particular entry crystallized her conflicting thoughts about how she could ever manage to push forward absent her life partner: "how many times did you say to me yourself that 'if you didn't have me, you might work, but you would be nothing more than a body without a soul.' And how will I find a soul when mine has left with you?"[45]

The bond that had fused the genius of Pierre and Marie Curie had provided a guiding light without which the world was truly a place she now felt held no meaning. But, in due course, Marie was to find that the pursuit of scientific discovery, so much an inherent part of what was the essence of Pierre Curie, along with the presence of two loving daughters from their union, would have to be enough to embrace and carry her through her darkest of moments as she fought to stabilize a world that now had no other focus. And, as only human experience can attest, time, the greatest salve for a broken heart, would slowly allow Marie Curie to resurrect a full life.

CHAPTER FOUR

The First Solvay Conference: Science at a Crossroads

It was called the Great Pestilence. During two distinct periods over the course of 1,100 years, from the 6th to the 17th century, it ravaged Europe and Eurasia. It first appeared in the Eastern Roman Empire in the mid-500s, recurring intermittently until the mid-700s. Then it vanished until its resurgence in the mid-1300s with outbreaks occurring periodically thereafter until the early 1700s. It emanated from farther east, brought by ship or caravan, leaving behind death so hideous and incredible in number as to overwhelm and decimate the terrified population, though a significant portion of which was, by physical immunities or good fortune, able to survive it.

In its most familiar form, as the bubonic plague, it invaded through the bite of human flesh from a black rat's flea, the insect injecting its host with a bacterium that nestled in the body's lymph nodes, especially in the neck, under the armpits, and in the groin. This caused pronounced and painful swollen glands called buboes, from which the term "bubonic" was derived. The buboes usually turned black, often accompanied by dark splotches or rings on other parts of the body, leading to the disease eventually being branded the "Black Death." If the lymphatic system poisoned the bloodstream, the bacterial infection could travel to any part of the body and toxify it. If it infected the lungs, it could be spread from the host's body to another through airborne droplets propelled by coughing and sneezing. Most of the affected lasted only a handful of days, with mortality dreadfully high at rates approaching 50 percent, death arriving only after being tortured by spiking fever, grievously aching muscles, often delirium, and vomiting of blood.

All this would later be classified simply as the plague. In England it last appeared during the 1600s, usually in London and then spreading to other towns and cities with almost alarming regularity and varying mortality rates every fifteen to twenty years from 1603 onward, rearing its ugly head for its last horrible surge of death in 1665–1666.[1] This last epidemic would become known as the Great Plague of London in which thousands died each week, approaching 100,000 in total, approximately a quarter of the city's population.

At the time of London's Great Plague, Cambridge was a town sixty-five miles north that housed a number of colleges founded over the years under the umbrella of the university. In the spring of 1665, rumors of the beginnings of another round of plague in London had already spread to Cambridge, sending a fearful shiver through the community as its elders recalled previous devastating encounters with the disease from years gone by. By midsummer, in July, the first of its inhabitants died from the plague, a young boy tattooed with the telltale dark rings, soon followed by his even younger brother, similarly adorned.[2] The deadly scourge had arrived.

Upon hearing of the spreading pestilence, the university took its customary social-distancing measure when confronted with the plague by sending its students fleeing to the countryside and shutting down. One of those sent packing was a student who had just received his undergraduate degree at Trinity College and was embarking on a masters, Isaac Newton. He lived sixty miles northwest, at a small, relatively isolated manor in Woolsthorpe, Lincolnshire. There, for most of the next eighteen months, Newton spent time in his home or walking the gardens and fields, mostly thinking and experimenting. What developed in that period within a mind not previously noted for brilliance was, in fact, the beginnings of what could fairly be called scientific genius, the likes of which would not be seen again for over two centuries.

From an early age, Newton was interested in the physical world. He was small in stature and an introvert, lacking any desire to play with boys his age who were mostly larger and more aggressive. Rather, he was a keen observer of his surroundings. When he was a child, Newton took great pleasure in building sturdy wooden models of things that people used in everyday life. He first fashioned small pieces of furniture, tables, chairs, and cupboards. Then, on to wheels, carts, and more creative items such as kites, sundials, and waterwheels, devices that combined natural forces like wind, sunshine, and water with practical purpose. Newton's curiosity concerning how the physical world worked appeared to stimulate in him a desire to understand how some man-made machines could function in concert with basic elements of nature.

As he grew, Newton turned his eye toward the heavens, expanding his investigations to try to decipher the relationships of the sun, moon, planets, and stars that filled the sky. Questions concerning the composition of sunlight, the interplay of Earth with the sun and other heavenly bodies, and the potential that all of this could be explained by a set of universal principles proven with mathematic precision were to drive Newton's inquisitiveness for many years. While on his extended respite from Cambridge during the plague, apparently he focused with a burst of intensity during the copious amounts of time that he had to devote to this type of reflection. He had read a great deal in the previous few years attending Trinity College before the plague, investigating a combination of mathematical, optics, and physics treatises, especially works by Descartes, Galileo, and Kepler. Toward the end of his life, having become the toast of the scientific world for his brilliant insights, Newton was to reflect on his unsurpassed inventiveness in these disciplines during that time by noting all this "was in the two plague years of 1665 and 1666, for in those days I was in the prime of my age for invention & minded Mathematicks and Philosophy more than at any time since."[3] Although he did not develop the full breadth of his theories on physics and mathematics all within these eighteen months, Newton certainly felt these were by far the most productive days of his scientific genius at work. In fact, it was 1666 that has been specifically noted by others as his "annus mirabilis," his "miraculous year," for the scientific explorations he undertook and the accomplishments that resulted.

Newton's efforts during this period were truly amazing. His study of sunlight, much of it conducted in the privacy of his own room with the aid of a few prisms and a pinhole he drilled in his window's shutter blind, yielded the result that white light was actually made up of colored components that spanned the hues of the rainbow. This in itself was not a unique observation, but his experimentation gave conclusive proof of the unequal degree of refraction of different colors through the prism as to why this phenomenon was so.[4] He speculated that light was made up of moving corpuscles, or tiny particles, rather than waves, which was the accepted theory of the time. But he did not experimentally validate this hypothesis, unlike most all of his other investigations where he conducted detailed experiments that helped him center on provable theories. His learnings about light allowed him to proceed with this knowledge to design a novel reflecting telescope, a major improvement that employed mirrors rather than refraction lenses used in existing devices, which caused chromic color interference while viewing through the device. These optical findings combined to bring Newton an invitation to join the Royal Society of London in 1672, long before his mastery of physics and mathematics was displayed in his 1687 groundbreaking publication *The Mathematical Principles of Natural Philosophy*.

Newton was a diligent experimental scientist, to be sure, but in a few instances took his investigations well beyond what any individual would deem prudent. Such was one case during his optics work when he famously chose to conduct a bit of self-experimentation that included "inserting a blunt needle, or bodkin, 'betwixt my eye & ye bone as neare to ye backside of my eye as I could' [in the eye socket] in order to alter the curvature of the retina [at the back of the eye] to observe the colored circles that appeared as he depressed the bodkin."[5] It seemed from these manipulations that colors could be seen by manually changing the shape of the eyeball itself, but this experiment with the bodkin certainly was an extreme demonstration of devotion to the search for firsthand knowledge. In general, his adherence to amassing experimental evidence to support a theory, rather than proposing hypotheses without data to back them up, was essentially the creation of what would become known as the scientific method, utilized to this day.

Bookending Newton's optical investigations during those few years at Woolsthorpe were studies of mathematics that eventually resulted in his creation of what he called "the direct and inverse fluxions," a convoluted name for the equally complex mathematical concept of calculus. Newton's intense interest in understanding the heavens and the movement of the sun, planets, and stars, and his devotion to proving his theories about these interactions as well as those between any physical body and another, almost inevitably led Newton to develop his work on fluxions. It was a mathematical system that could be employed in describing how the motion of all of the bodies in the universe were related.

The fluxions, what came to be known as differential and integral calculus, are used to analyze rates of "changing quantities"—especially "bodies in motion," like the change in velocity of an object as time passes, the decelerating speed of an apple tossed up in the air, or the motion of the moon circling Earth.[6] A practical example of how calculus is used relates to NASA engineers designing a rocket that propels itself into space. In launching a rocket, it is vital to know how much thrust, or force, is needed to move it through the air such that it can overcome Earth's gravitational pull in the atmosphere and reach outer space. By employing calculus, the engineers can determine the thrust needed to accomplish this by examining the acceleration of the rocket after ignition, calculating the rate of change of the velocity of the rocket, and factoring in the mass of the rocket, which is a continuously decreasing quantity as rocket fuel is burned and stages of the rocket are jettisoned from the vehicle assembly.

Most everyone is acquainted with the story of Newton seeing an apple fall from a tree to the ground below in his Woolsthorpe garden. Some fanciful versions of the event even depict him lying under the tree when the apple fell from a branch and supposedly hit him on the head, stimulating an idea and perhaps a headache as well. Whether apocryphal or not, the point was that this very simple example encompasses the greatness of a concept that, like calculus, can broadly be applied to any two bodies in the physical world: gravity. Newton's observation of the apple falling to the ground is a metaphor for how two physical bodies in the real world are attracted to each other, the overriding phenomenon of gravity that was to enshrine Newton in the annals of history.

Newton used the example of the apple to incrementally expand his understanding of the phenomenon of gravity to encompass apples higher up in the tree, then to apples in even higher trees, all the way to wondering why this same effect would not exist between bodies in the heavens, like the moon and Earth. He saw that the apple fell straight down from the tree branch to the ground below, not off at an angle, from this observation determining that the attractive force of Earth was at its center, not off to one side or the other, signifying that gravity emanated from the physical center of each body that exerted it. He determined that this gravitational force was directly related to the mass of the object, the bigger the mass the larger the gravitational force exerted by the body on other bodies. And he noted that the closer the body was to other bodies, the stronger the gravitational force was, such that bodies farther away from each other exerted less force on each other than those that were closer together. When the framework of these observations was shifted from an apple falling to the ground to one that included the moon and Earth or the sun and the planets, the phenomenon of gravity was equally applicable, helping to explain the motion of all sorts of heavenly bodies in relation to each other and, with the help of calculus, mathematically determining their paths of movement.

Newton was to codify all of this in 1687 in what is arguably the greatest single scientific publication of all time, *The Mathematical Principles of Natural Philosophy*. From ancient times, "natural philosophy" was a term meant to convey the study of nature, the physical world. Newton's book title extended this study to include the use of mathematics to describe nature, with it giving modern science precious tools to prove its theories and a language to convey them. In addition to optics and gravity, Newton also expounded upon three laws of motion that have become equally famous. The first law, known as the law of inertia, commonly states that "a body at rest tends to stay at rest," which means that unless some sort of force is exerted on the body to make it move, it simply stays put. The second law notes that if a body of a certain mass is accelerating in motion, it develops a force that increases as it accelerates. An example of this would be a compact car that is accelerating. If it hits a wall, it hits the wall with a force that is directly related to the weight of the car and the acceleration of the vehicle. If the car in the example is replaced by a large truck with the exact same acceleration, the force with which it hits the wall is greater than that generated by the car, simply because the truck's mass is greater than that of the car. The third law is probably the best known; it says that "for every action there is an equal and opposite reaction." This is demonstrated when someone shoots a gun. In the instant the gun launches a bullet, sending the projectile out of the gun's muzzle at a certain velocity in a specific direction, the shooter experiences the recoiling force of the gun, moving in the exact opposite direction. These laws, in conjunction with the phenomenon of gravity, established principles by which scientists for the next two centuries could accurately and in precise detail describe much of how the visible, physical universe worked.

With these concepts, Newton could use mathematical proofs to verify Kepler's hypothesis that was stated a generation before, that planets revolved around the sun along elliptical paths, as opposed to the notion of Copernicus that planets etched perfectly circular orbits. Others were coming to believe Kepler's idea but were unable to prove it with verifiable data until Newton came along. In another instance, the ocean tides' ebb and flow could be explained by the moon's gravitational attraction to Earth. Even 150 years later, Newton's laws and theories were proven to be sound. In 1781, a new planet, Uranus, was identified in the night sky by astronomers. However, over the next fifty years it was observed that the orbit of Uranus around the sun varied enough from Newton's calculations to cause the scientific community some concern. That is, until the application of Newton's gravitational theories predicted that another large mass, big enough to be another planet, was causing the aberration. By using Newton's predictive calculus, the exact location of this new planet, Neptune, was established and identified.[7] Here was Newton's masterpiece, a "philosophy of nature based on the principle of forces" acting on each other, explaining the physical phenomena on Earth and, indeed, the wonders of the solar system based on gravity.[8]

These theories covered one basic portion of the visible, physical world, everything composed of matter. The other portion of the world, radiation or energy, was in part addressed by Newton's studies of light as a form of radiation. His mathematical description of light as composed of individual colored rays that could be precisely described by their measurable, differing angles of refraction through a prism, clearly demonstrated the power of his mathematical characterization of energy.

John Maynard Keynes, famed economist as well as Newton devotee, "believed that the clue to Newton's mind was in his unusual powers of continuous concentration, and that he 'could hold a problem in his mind for hours and days and weeks until it surrendered to him its secret.'"[9] Of course, during the months Newton was confined to his Woolsthorpe manor due to the plague, he really had nothing but time to devote to examination of his questions of the workings of the world, applying his intense mental focus to maximum effect. That he was able to develop such astounding insight into the relationships of physical bodies in the universe and their mechanics, as well as devise a mathematical system to represent it all, certainly qualified as the pinnacle of genius. Newton's theories were the result of careful observation and relentless experimentation, a process that was the only one that he found sensible. In his view the heretofore accepted practice of scientists spouting unproven hypotheses lacking in verification was the cause of poor conclusions. But how a mediocre student with no special training nor history of brilliance could revolutionize science and become the most famous mathematician and physicist in Europe was in some ways beyond comprehension, at least until it happened again in the person of another obscure individual at the beginning of the 20th century.

It took over fifty years for science to absorb, verify, and institutionalize all of what Newton's fertile mind had wrought. By the mid-1700s, what had become scientific gospel in England was generally accepted by most on the European continent as the incontrovertible view of how the physical world worked. White light was understood as energy made of constituent colors of differing refraction and matter was shown to follow distinct laws of motion, crowned by the principle of gravity and supported by mathematical proofs based on calculus.

Then, in the 1800s, the discovery that electricity and magnetism, both well known but little understood scientific curiosities for centuries, were inherently linked was demonstrated by the efforts of the self-educated Englishman Michael Faraday and his next-generation supporter, Cambridge physicist James Clerk Maxwell. Their work masterfully elucidated the fact that, simultaneous to the realm of visible matter governed by Newton's laws of motion and gravity, there existed throughout the world invisible phenomena such as electricity and magnetism as transparent but quite real fields of energy. As Nancy Forbes and Basil Mahon described it in their fascinating book on Faraday's and Maxwell's efforts, "space itself acted as a repository of energy and a transmitter of forces: it was home to something that pervades the physical world yet was inexplicable in Newtonian terms—the electromagnetic field."[10]

This energy traveled in waves of varying lengths, from those visible as the light that Newton had studied to periodically visible electricity (as sparks or lightning), invisible magnetism, and even as yet-to-be-identified radio waves and microwaves, which all traveled at the speed of light. Physicists proposed that these waves were carried by what they vaguely termed the "ether," an invisible theoretical construct that supposedly was present everywhere in space as the medium to support the spread of these waves, much like water facilitated the generation of waves that rippled outward in a still pond from a point in which a stone was dropped. Maxwell creatively applied Newton's differential calculus in the mid-1800s to mathematically describe the fields, but the concept was not proven experimentally until German Heinrich Hertz's electromagnetic investigations in 1888.

During the 1890s even more waves were being detected, including Röntgen's X-rays and the Curies' radioactivity. Still, the physics of the visible world of Newton, where forces acted on each other from varying distances, and the invisible fields of Faraday and Maxwell maintained a peaceful coexistence in the scientific community. At the same time, the intriguing detection of particles smaller than the atom itself was taking place, specifically the electron proposed by Dutchman Hendrik Lorentz in 1896 and verified by Englishman J. J. Thomson in 1897 in his experimentation with cathode rays. This opened up the mysterious subatomic universe, one composed of constituents that would soon appear to have a very different way of acting than prescribed by classical Newtonian physics.

As the 20th century dawned, in 1900 the German physicist Max Planck tackled a theoretical conundrum that could only be explained by bending classical physics to the breaking point. Earlier in his life, in the mid-1870s, Planck was trying to decide between pursuing his promising potential as a pianist or obtaining a degree in physics. Somewhat surprisingly, he chose the latter in spite of one of his University of Munich professors, Philipp von Jolly, who discouragingly advised the young student that, in physics, "almost everything has been discovered and all that remains is to fill a few holes."[11] Still, Planck persisted in his desire to be a physicist, eventually becoming one of the best-known in Europe. He was conservative by nature, and strictly hewed to Newtonian principles until a focus of his physics experimentation led him to develop a radical theory that would be the basis of undermining a portion of the very foundations of classic physics.

Planck's special scientific interests centered on the study of thermodynamics, the relationship of heat to other forms of energy, like electrical or chemical. For many scientists of the era, a particular curiosity in this field related to what was called "black body radiation." This referred to the theoretical concept that some types of matter did not reflect any light, but only absorbed it. This lack of light reflection from the body caused it to be observed as black. Some theoretical experiments contemplated the spectrum of light that would be emitted from this black body, based on the heat of the energy released from light absorbed by it. Practical applications of this investigation were of help in determining things such as the energy contained in light emitted from lighting filaments, which approximated black bodies in function.

One of Planck's investigations on black body radiation of heat caused him to conclude that beyond behaving like a wave in the ether, under certain circumstances energy could behave as a series of discrete packets, or "quanta." Although a radical departure from accepted concepts of physics, which Planck himself was almost embarrassed to postulate, "the evidence for its validity gradually became overwhelming as its application accounted for many discrepancies between observed phenomena and classical theory."[12] Planck had reluctantly become the first of what would be a growing number of physicists who proposed this unique departure from classical physics, assisting in the development of what would later be called quantum theory. For many years, Planck appeared to be an enemy of his own proposition in this regard, routinely explaining that his was simply a theoretical solution developed for mathematical completeness of his investigations, being difficult, if not impossible, to prove with tangible examples. With this type of ambiguity, it was a wonder that any scientists at all thought it might be valid.

Yet, this marked what was to be the beginning of a major shift in scientific thinking of the past two hundred years, in which time Newtonian physics had become firmly established as dogma. Newton's laws of motion, gravitational theory, and understanding of light were the basic principles on which much of physics was supported. The work of Faraday and Maxwell, though addressing fields of energy that were not really understood at all in Newton's time, was also viewed as accepted theory that could exist in harmony with Newton's physics, employing explanations of these phenomena with Newtonian calculus. However, with investigations like Planck's newfound theory of energy quanta and Thomson's recent discovery of the subatomic electron particle, the foundation of what would later be termed "classical physics" was beginning to be chipped apart. Fissures in the neatly constructed and accepted theories of the natural world were becoming visible to a growing number of physicists across Europe.

A few years later in 1905, an innocuous twenty-six-year-old named Albert Einstein, a Swiss patent examiner not especially noteworthy for his training or intelligence, published a number of papers that would have profound implications for the classic Newtonian view of the world. A few of them distinctly supported Planck's findings, eventually becoming in large part the underpinnings of this new view that would soon shake classical physics to its core. This was a daunting proposition, as Einstein himself commented as he viewed in hindsight the physics landscape of the early 1900s in a short autobiographical piece published in 1949, "dogmatic rigidity prevailed in matters of principles: in the beginning (if there was such a thing) God created Newton's laws of motion together with the necessary masses and forces . . . light [was] the wave-motion of a quasi-rigid elastic ether . . . all physicists of the [19th] century saw in classic mechanics a firm and final foundation for all physics."[13]

The classical physics principles laid down by Newton over two centuries before had by this time become engrained, even revered, in the scientific community. Faraday, Maxwell, and experimental physicists had succeeded in completing this masterpiece by the addition of much-needed brushstrokes depicting the electromagnetic fields clearly in the background with their associated waves of energy. But, for the most part, the statement on the relative completeness of the science offered by Planck's university professor von Jolly twenty years before reverberated in most physicists' thinking. American scientist A. A. Michelson, in his address for the dedication of the Ryerson Physical Laboratory at the University of Chicago in 1894, touched on the state of physics experimentation in this manner: "it seems probable that most of the grand underlying principles have been firmly established and that further advances are to be sought chiefly in the rigorous application of these principles to all the phenomena which come under our notice . . . An eminent physicist has remarked that the future truths of Physical Science are to be looked for in the sixth place of decimals."[14] This was meant to be a proud statement of the grand state of physics and its relative completeness as a science. Yet, at the same time it was a decidedly sobering assessment of the situation, one that was shared by many in the field. The last sentence of Michelson's statement, referring to the importance of tiny differences in experimental results versus predicted theoretical expectations, surely did represent a gap where nuggets of scientific discovery still might be found. But the overall lack of excitement in this notion as it related to what was seemingly all that remained of the future of scientific research was deflating, especially since supposedly the "eminent physicist" to whom this remark was attributed was none other than a towering figure in the scientific world, the deeply respected Irish-Scottish scientist Lord Kelvin.

Now, Planck's unintentional discovery of a theoretical solution to a hypothetical thermodynamics problem had become the opening act eventually leading to an inevitable finale that Thomas Kuhn, modern American physicist and philosopher, would call a bona fide scientific upheaval. In his seminal text on the history of change in science, Kuhn described this type of change as a time when "normal," accepted tenets of scientific thought are disrupted by anomalies that can't be solved by established theory. As the discrepancies begin to pile up, the sheer weight of their unanswered questions generates a crisis in the stability of the discipline that can only be addressed by the assumption of new scientific principles.[15] As Planck and others discovered greater discontinuities that they could not explain through "normal" science, the crisis deepened, seeking a solution.

By the end of the decade, more than a few European scientists were beginning to question the neatly tailored theories of Newtonian physics, which expertly explained the mechanical forces of the physical world of visible bodies. More experimental results were leading to conclusions that in some cases could not be calculated through use of Newton's accepted tools or even Maxwell's equations, especially related to some forms of energy waves and subatomic particles. Einstein's postulations in his papers of 1905 were instrumental in furthering the debate on Planck's energy quanta theory, as well as addressing other physics phenomena that had previously gone unanswered with classic principles. These efforts would add to the developing surge of evidence being compiled, all pointing toward the need to address the mounting discrepancies and somehow adjust the scientific view of the natural universe.

To this end, in 1910 German physicists Planck and Walther Nernst exchanged written correspondence concerning what they "perceived as a no longer tenable set of contradictions and difficulties in physics,"[16] expressing varying degrees of alarm about the situation. Planck had stumbled upon his theory, repeatedly proven by experimentation, of the duality of energy's existence as a wave and as packets of energy in special cases. But few physicists were aware of this at the time, and those who were aware didn't know whether this was indeed a significant finding or a curious anomaly. Independently, Nernst was arriving at results based on low-temperature observations of matter that were supportive of Einstein's proposed ideas about quantum theory. Nernst felt his low-temperature research was groundbreaking and in fact worthy of great honor, perhaps even a Nobel Prize, in keeping with his high opinion of himself and his efforts.

"Walther Nernst was very ambitious," as Franklin Lambert, historian and physics professor at the University of Brussels, described it in an article on the subject. "He wanted the Nobel Prize. But he knew that, for this, quantum theory, at that stage very controversial, had to be validated at the highest level."[17] To accomplish this, Nernst thought it was now time to convene a group to deal with the issue, to acquaint a larger portion of the scientific elite on the subject, discuss various viewpoints, and try to collectively arrive at some sort of consensus in addressing the situation. Planck was more cautious, ever the traditionalist, countering with the view that more experimentation was required before setting up a gathering that might result in a potential rush to judgment on the subject. In fact, he suggested to Nernst that besides the two of them, only a handful of other physicists in Europe (two of which were Einstein and Lorentz) would currently be interested in a meeting to discuss the situation.[18]

Undeterred, Nernst approached the wealthy Belgian industrialist Ernest Solvay with the idea of sponsoring a scientific conference unlike anything seen before. Solvay was a well-known philanthropist and liberal mind, having long ago turned the reins of his chemical business empire over to others in his family while he focused on funding ideas that might improve society. He had already established Belgian institutes of higher learning on medicine and sociology, the Institute of Physiology in 1895 and the Institute of Sociology in 1902, as well as funded the establishment of a University of Brussels department in commerce, and engaged numerous Belgian scientists in these organizations.[19] Nernst suggested bringing together a handpicked cadre of European physicists from half a dozen countries, have them present their views on the puzzles of energy and the quanta that currently faced the discipline, initiate discussion, and see where it all would lead. Solvay was generally agreeable to the concept, especially since it would allow him, as sponsor, to offer some suggestions of his own related to scientific theory; he was a creatively enthusiastic, if eccentric, thinker.[20] He had harbored a passion for physics and chemistry from an early age and, even though self-taught in each, was an innovative promoter of both. The gathering would be termed a council, unusual for a scientific meeting, but actually more appropriate for Solvay, who viewed it all as "a sort of private conference," what might be termed a council of wise individuals, a panel advisory to himself in his search for scientific truths.[21]

The sessions were completely underwritten by Solvay, who offered each participant one thousand Belgian francs to cover travel, lodging, and dining expenses. The conference would take place at the Grand Hotel Metropole in Brussels, an ornate establishment built by the Wielemans brewing family to augment their adjacent, tremendously successful café that featured their beers.[22] In the mid-1890s, the hotel had been converted from a neighboring bank building into what would become the most modern travel lodging in Belgium, rivaling any in Europe, employing the new commercially available offerings of electricity and central heating for the supreme comfort of their guests. The building's design was an amalgamation of numerous architectural types, including gothic, classic French Renaissance, and Art Nouveau, the latest rage in artistic style. The grand entryway led to a columned lobby showcasing gold-tinted stained-glass windows and glittering chandeliers, the sumptuousness of the surroundings conveyed through the liberal use of luxurious marbles, mirrors, gilded bronze, and exotic teakwood.[23] Art Nouveau had supplanted the more formal architecture of the mid-1800s with the sweeping curves, natural patterns, and rich materials that were trademarks of the movement, breaking with the stiffer, more formal motifs that had been generated from the industrialization of the world earlier in the century and its subsequent expression in art.

The Hotel Metropole had wonderfully blended old and new architectural concepts into a complementary feast for the senses. Similarly, a number of the scientists attending the conference hosted by Solvay at this venue hoped that the group might come to an elegant compromise of classic and new theories of physics at this time of potentially disorienting theoretical change. Instead, what more than a few felt resulted was as much a wistful yearning for the continuation of the greatness of Newton and Maxwell as it was the begrudging acceptance that neither's mighty theories could explain the new mysteries unfolding before their eyes. As professor of history John Heilbron noted about the conference in an address at a more recent Solvay gathering, "Most of the participants took away the conviction that the foundations of physics had to be enlarged, if not built anew, to accommodate the quantum of energy unintentionally introduced into the theory of radiation by Max Planck a decade earlier."[24]

Surprisingly, Einstein gave a rather harsh summation of his feeling on the gathering in a letter to a close friend shortly after the meeting. "In general, the congress in Brussels resembled the lamentation on the ruins of Jerusalem . . . Nothing positive has come out of it . . . I did not find it very stimulating, because I heard nothing that I had not known before."[25] Though nothing new may have revealed itself to Einstein, this was perhaps because much of what was reviewed at the gathering were thoughts he himself had expressed over the past few years. But the central ideas being discussed were stark revelations to many in attendance, evidence that could not be ignored as to specific failings of classical physics theory. Moreover, what Einstein failed to incorporate into his thinking about the conference was that, as much as he was meeting many of the exalted royalty of the European physics world in Brussels for the first time to discuss their collective scientific future, the others were taking the measure of him as well. Planck, Nernst, and Lorentz were more than acquaintances, having met Einstein at previous scientific gatherings over the past two years. But many of the others, including Marie Curie and the four other physicists of the French contingent at the Solvay Council, were observing Einstein's brilliance in person for the first time. In doing so, a number of influential participants began to devise the professional calculus for the launch of Einstein's career, which would eventually rocket his theoretical genius into the furthest reaches of space and time.

CHAPTER FIVE

Solvay's Search for a Universal Law

In 1905, energy and matter were determined to have a direct relationship by the great theoretical genius of Albert Einstein. Nine years before this astounding pronouncement, in a little known but even more surprising development, an independently wealthy Belgian chemical entrepreneur stated similar conclusions that he had forwarded to the Belgian Academy of Sciences. In them he noted in part, "I have gained the conviction that matter is transformable to energy and *vice versa*... One of these elements is, so to speak, only a form, a modality or, more correctly, a transformation of the other. My ideas on this subject go back to 1858."[1] In addition to being a master businessman, Belgian Ernest Solvay was a self-taught scientist with no university education due to an extended bout with pleurisy as a teenager that had kept him homebound. He had initially speculated on this conclusion about matter and energy when he was teaching himself about physics and chemistry while employed by his uncle in the latter's small chemical and gas works. Solvay was to pursue the experimental verification of his intuitive conclusion on this equivalency for over fifty years, the proof eluding him in large part due to his lack of formal knowledge of theoretical physics as well as its associated experimental techniques. But, when the opportunity presented itself to discuss his ideas, in the form of Walther Nernst's suggestion of Solvay's sponsorship of a select gathering of European scientific elites in 1911, Solvay couldn't help but relish the prospect.

Ernest Solvay was a daring thinker. His first and boldest experiment was that of taking a previously known but almost impossible to run chemical process and making it work efficiently. A vital product for the newly industrializing world, sodium carbonate, more commonly called soda ash or just "soda," had previously been made by the Leblanc process, which produced soda of varying purity, at high cost and with unwanted, polluting by-products. By contrast, Solvay's proposed innovation would make high-purity soda with significantly lower production costs and much less pollution.

Soda was one of the most important chemical additives or processing aids employed as the world initially industrialized during the 19th century. It was a much-needed ingredient in the manufacture of glass, soap, textiles, and paper, to name some of its more important uses. In making glass, addition of soda helped to reduce the production temperature needed to melt silica (sand), the major raw material of the process. This translated to lower energy costs, making soda a sought-after additive. In soaps and detergents, soda was important in helping to remove stains and to soften hard water. For textiles, soda helped natural dyes adhere to fibers. And the list of its uses went on and on.

To produce soda efficiently and in large quantities was more complex than it appeared just from its laboratory production route employing some basic chemicals. Solvay happened upon his improvements in making soda while experimenting in finding uses for excess ammonia gas being produced in his uncle's chemical plant. He intuitively knew his improved process would be a game-changer for the industry, with a fortune to be made if it could be successfully run. Solvay quickly filed a patent in Belgium in 1861 for making soda with three simple ingredients: ammonia gas, salt from brine, and limestone. The following year, he even thought he might quickly turn a profit by selling the rights to this patent to another Belgian chemical producer. This company would use it to improve their in-house production of soda they currently made via the expensive, polluting Leblanc process for use in their own manufacturing of glass.[2] Unfortunately, Solvay's filing a patent in Belgium only meant that the idea was officially recorded. In a disheartening turn of events, the patent sale fell through when a recently met company lawyer, advisor, and soon to be partner, Eudore Pirmez, quickly determined that the patent on Solvay's combination of these products to make soda was invalid. Unbeknown to Solvay when he filed his patent, a dozen others in Europe had tried over the past fifty years to use the exact same materials to make soda, some of them filing patents well before him. Each had similar problems with the process: making a few pounds in the laboratory was easy, but producing soda using these same ingredients on a larger, industrial manufacturing scale of thousands of pounds while achieving high purity was difficult and costly.

Solvay was not to be deterred. In the early 1860s, Ernest enlisted his younger brother Alfred to the cause and formed a company that was dedicated to delivering large volumes of high-purity soda. After running a small experimental plant to prove the process could be run on larger than a laboratory scale, they established their business in 1863. Over the next few years they began to make larger volumes of soda but suffered a series of setbacks that drove the tiny enterprise to the brink of bankruptcy. Equipment failures, leaks, and breakdowns from running the process improperly left the manufacturing in almost a continual state of disrepair. The vessel used to mix and react the raw materials would intermittently clog up and halt the process entirely.[3]

The brothers plowed ahead, seeing glimpses of hope on days when the manufacturing facility ran smoothly and hundreds of pounds of soda were made, only to be followed by more equipment problems that halted production altogether. Finally, in the fall of 1865, an explosion rendered their primary reactor unusable, seeming to end their dreams of success. Now at their lowest, the brothers decided it was time to hold a family council to determine if they should proceed further or just give up. The brothers' parents, who had made initial investments in the business along with some other family members and a few close friends, signaled unwavering belief in the vision of their two cherished sons by offering a precious forty thousand francs of their savings as a final injection of capital into the venture.[4]

This support was invaluable, and with it the Solvay brothers regained their determination to succeed. Now, promising production results ensued. By February 1866, they were making thousands of pounds of soda each day, clearly indicating their efforts were finally beginning to bear the sweet fruits of their labor. At the end of the month, Ernest was to write of this remarkable turnaround, "After being so often within an inch, within a hairs-breadth of failure, can I finally succeed in regaining from others and from myself, that moral strength which the entire world seems to refuse me. After all that has happened, this would indeed be a splendid triumph! Let us not speak of it, we are too accustomed to setbacks and deceptions."[5] The worst was over, with a seemingly boundless future for the Solvay brothers if they could manage to consolidate their gains and keep the process running.

Of course, this dramatic change in fortune didn't happen only due to a timely infusion of monetary support accompanied by fervent prayer, though both helped. Ernest had continuously labored over the failures of the equipment producing soda. His self-schooled knowledge of chemistry, combined with on-the-job training in his family's gas operations, had honed a razor-sharp, problem-solving mind that could focus on the mechanical as well as the chemical production issues confronting their efforts and creatively imagine ways to improve the process. Solvay was innovative in his approach, not afraid to significantly adjust the process to see if he could improve it. After all, with their previous failures now leading to the very precipice of insolvency, what was there to lose in these attempts? Granted, he was not trained as a chemist, whose primary focus would be to discover new chemicals as well as novel paths to make them. Nor was he a degreed engineer, whose mission would be to design, build, or maintain machines. But Solvay's activities were centered on working shoulder to shoulder with a few key employees to improve each major step of the soda manufacturing process. As professor of chemical engineering Laurence Scriven described it, Solvay and his team closely examined the process and "[divided] it into distinct operations . . . inventing new types of equipment for [making product] continuously on a large scale" versus making soda in many discrete, smaller batches.[6] Solvay and team determined the best temperature conditions and ideal mixing methods for the raw materials to react and decided on how best to get the highest purity for the final production of soda. In large part, this was the essence of what would later come to define the practice of "chemical engineering" before it was even a term associated with this type of activity.[7]

By the early 1870s, Ernest was at his chemical engineering best when he helped develop and patent a piece of equipment that allowed the process ingredients to intimately mix and react with each other more completely. It would become known as the Solvay tower, an imposing seventy-five-foot-tall column that improved manufacturing efficiencies, helping to increase the production of the process while lowering costs. It turned out to be one of the critical factors in distinguishing the company above all other soda producers.[8] In fact, under the guidance of attorney Pirmez, Solvay now forged a different patenting strategy employing these novel equipment designs. Beyond patenting the chemical process for the ingredients that were combined to make soda, which he and others had already done with nothing to show for it, Solvay proceeded to patent vital pieces of equipment and the variables involved in making them work optimally. In total, "A new strategy ensued . . . patenting everything, everywhere . . . a patent for each stage of technical operation and each new device required for manufacture at the industrial scale."[9] Solvay developed a defensible set of worldwide patents, eventually offering this technology to a select few in exchange for licensing payments for use in making soda.

At the same time, the demand for low-cost, high-purity soda to expand on or replace existing production versus the old Leblanc method of soda manufacture was accelerating at a tremendous pace. Over the next twenty years, Solvay branched out to build soda factories in other European countries outside of Belgium, including France, Germany, Austria-Hungary, and Italy. He established joint soda ventures with strong business entities in key international geographies to which he licensed the patented technology, especially in England, Russia, and America. Taken as a whole, this approach heralded the eventual death knell for competitors employing the LeBlanc process while producing riches for Solvay and his small circle of company investors that were beyond their wildest expectations. As the company dominated the international soda marketplace and began to make many other industrial chemicals, by the beginning of World War I Solvay's firm was the largest chemical company in the world.

As his company grew in size, employing more workers in production plants throughout Europe, Solvay's basic humanistic values permeated the organization in various ways. These values gave rise to a paternalistic model for senior management to institute that was developed for their employees in order to establish stability and general satisfaction in their work lives, hoping in return to engender loyalty and dedication to their jobs. More broadly, social support programs were beginning to emerge on a national level in various European countries during the 1880s. Chancellor Otto von Bismarck of the recently unified German federation of formerly independent principalities was among the first to establish nationally coordinated social welfare policy. Politically, for Bismarck it was an important aspect of developing and retaining loyalty to the new German nation from the people of various former states. A series of policies supporting medical treatment and sick leave, worker's accident compensation insurance, and old age pension programs was put in place, financed by a various combination of employer, employee, and government contributions.

Some progressive leaders in European industry viewed this effort as a potential blueprint for developing worker loyalty and dedication during a time of major upheaval in employment as society moved from a primarily agrarian focus to newly developing industrial enterprises. Solvay's business enthusiastically supported this path, blending the founder's embrace of a family approach to business and empathy toward the burdens of industrial workers with the desire to improve workplace productivity. In fact, by anticipating the inevitability of the institution of government-sponsored social programs for workers across Europe, Solvay's own leading-edge employee approach ensured relatively smooth operations within its facilities through establishment of a dedicated, generally contented workforce.[10]

The Solvay organization was one of the most progressive firms in this regard in most every country where it produced its materials.[11] Introduction of eight-hour workdays, pension funds, health insurance, and paid annual vacation time were some of the many programs instituted for the benefit of the workforce. With many of the production facilities throughout Europe far removed from urban areas, Solvay established company towns close to their manufacturing operations, offering affordable housing, primary schools, day care for working parents' children, and even evening classes for workers to learn mathematics, language, and history.[12] These programs varied somewhat by geographic location and cultural standards, but wherever this approach was introduced the effect was positive in strengthening the ties of the workers to the company.

Ernest Solvay was now a very wealthy man, having achieved what many thought was impossible. He had taken a chemical process deemed commercially unworkable and through tenacity, inventiveness, and at times sheer willpower, succeeded in developing a company that was the leader in its field. He had defied all expectations of his home-schooled abilities, literally creating the job description of what was to become the profession of chemical engineering, as much out of necessity for his firm's survival as from his supreme confidence in a vision of what was possible for his treasured soda ash process. He was known to many as a pioneer of the industrial chemical business, through innovation in the operations of his facilities and care for his workers. For the most part, this was a natural extension of his character as defined by his steadfast devotion to family, education, hard work, and a willingness, some might say an imperative, to explore the limits of human capabilities. The American chemist, businessman, and contemporary William Nichols, shortly after Solvay's death, described him as "simple, courageous, studious, affectionate, honest, and thoroughly human . . . the words of his father he had made his own: 'Work is a debt that all true citizens owe to society.'"[13]

By the closing decades of the 19th century, Solvay's financial success was established beyond anything he might have imagined. With forethought and planning, he then proceeded to pass the reins of daily control of his now-sprawling business empire to others in his limited group of family members and investors, especially his brother and trusted confidante Alfred, who had shared equally in the pains of birthing a new business as well as the rewards of helping it grow and prosper. He succeeded Ernest in running the company, but unfortunately was only to live another decade before an early death at fifty-four, delivering a personal blow to Ernest that was difficult to withstand. The siblings had been through so much together as they triumphed in their business venture against tremendous odds. For Alfred to leave this world at such an early age was a shock to his brother, who had begun adulthood with a frail constitution due to his bout with pleurisy. By this time Ernest's son Armand had completed his university education as an engineer and was old enough to become a part of the small executive committee that ran the company, extending active family leadership of the thriving enterprise into the 20th century.

Solvay was now free to concentrate on his true inner passion, the study of science and the exploration of his long-held intuitive theories. He sometimes explained that his efforts in industry had only been a means to an end, success with his company providing him with the financial independence to devote himself to science, to "create a scientific institute designed to affirm or contradict" his ideas.[14] Now, until his death, he would spend a major portion of his time and much of his personal fortune on contemplation, experimentation, and philanthropy regarding what he felt were science's unerring, logical solutions to physical phenomena and how they might be shown to encompass a better path forward for humanity. Solvay lovingly called this the devotion to his "fifth child to raise," beyond his own two sons and two daughters.[15] It was a tantalizing, complex effort that took Solvay down many paths, some more fruitful than others. In spite of all his industrial and business success and its accompanying princely trappings, deep inside was a yearning to leave a stronger mark on the world. As he described it, "To be in contact with scientists, to become in some small way a scientist myself if possible, perhaps to cast a new light on physical phenomena, to be able to uncover what is real and definitive, was my life's greatest dream."[16]

Ever since his plans of university study had been thwarted by his physical ailment, Solvay had thrown himself with conviction into his own personal education in science. He read voraciously, especially texts on physics and chemistry. These learnings were invaluable in the perfecting of an economically efficient soda process. More important, they would engender within Solvay an almost mystical belief in the laws of science as they pertained to outlining and guiding the workings of the physical world. As Newton had searched for answers in applying the phenomenon of gravity to make sense of the paths of heavenly bodies, so Solvay began to wonder if there was an underlying natural scientific principle that, if applied appropriately, might connect physics, chemistry, and the biological workings of the human body, intertwining to ultimately dictate the functioning of a more perfect society. He came to firmly believe that there must exist a "universal law" that governed these sciences, to the benefit of humanity if applied properly.[17]

To this end, a number of Solvay's thoughts and scientific investigations led him to embrace what was called the energetics movement of the late 1800s, a tangential theory of human interaction in society that appealed to his extreme devotion to rational thought and rigid systems. As the name of the discipline implied, it centered on energy as the main driving force of humanity, its sources and uses, and how it could be harnessed in the most practical, efficient manner to optimize an individual's contribution to society as well as maximize his or her own well-being. Further, Solvay held a rather eccentric opinion that "the progression of the universe was not attributed to either a vital force or matter endowed with a soul, but rather to energy transformation."[18] In this regard, he viewed the human body as a machine that converted energy from an input to the body, with the output yielding something beneficial to society, in doing so providing well-being for the individual completing the transformation.

The idea itself never caught on with more than a small group of dedicated disciples of the concept, being rejected by numerous sociologists and psychologists who questioned his insistence on trying to reduce human activity to, in effect, a series of mathematical equations. As an example, diet and nutrition were viewed as providing chemical "inputs" for the human machine, being a major factor in determining the quality of what that person could generate as "output" for the benefit of themselves as well as society. Practically speaking, those working on an empty stomach could testify only too well to the negative effects to be expected on their resulting accomplishments, or lack of, in the schoolroom or the workplace. But to try to quantify the absolute contribution that could be generated from dietary input of various foods and liquids and the resulting value of the outputs of the body, especially related to intellectual outputs like writing a symphony or a novel versus physical tasks of manual labor, was difficult if not impossible to objectively estimate.

Nevertheless, the logic Solvay employed in examining the optimization of human efforts in this manner, both individually and for a society, was quite useful in helping him to develop social proposals for the Belgian government during the years he served as a national senator during the 1890s. These included Solvay's arguing for "vocational training for the unemployed [as well as urging] the passage of an inheritance tax" meant to reduce the deposit of large fortunes benefitting nonproductive heirs, both concepts intended to adjust the inputs being provided to individuals in order to increase outputs to society's benefit.[19] Although not embraced even by many forward-thinkers of the day, both ideas have proceeded to gradually gain varying degrees of acceptance in cultures searching for a means to level society's playing field for individuals of differing backgrounds.

Among others, German chemist and future Nobel laureate Wilhelm Ostwald was a fellow energeticist. Solvay and Ostwald became acquainted over discussions on the concept well after both had individually become convinced of its utility, though Ostwald didn't share the more extreme opinions of Solvay on the subject. The two spent many hours together discussing this theory, its application, and the general desire to move society forward with their potential assistance through employing an energetic approach. Ostwald visited Solvay on numerous occasions, spending time at the latter's sumptuous estate outside of Brussels. In 1913, he was even invited to this manor for the celebration of the fiftieth anniversary of the founding of the Solvay company, which conveniently coincided with Solvay's golden wedding anniversary as well as an early marking of Solvay's seventy-fifth birthday, which actually fell in the following year. Taken all together, this triumvirate of important milestones made for a grand occasion. Ostwald, in his autobiography, notes Solvay's enduring connection with his workers by relating that at the celebration, Solvay had expressly prohibited personal gifts from the many directors of the company's international operations as well as senior management personnel attending the soiree. "Instead, he suggested that any money set aside for that purpose should be used to support the workers in the factories. It pleased him a lot that three million francs were raised for this purpose."[20]

Both friends wrote consistently on energetics, Ostwald even authoring a text entitled *Energetic Basis of Cultural Science,* which was published in 1908 and was dedicated to Solvay, whom he called "the founder of social energetics."[21] The book, as well as many an article penned by either man, was roundly criticized by the respected German sociologist Max Weber and other cultural experts as well-intentioned but missing the mark as a believable set of theories that could be employed to guide society. As Ostwald described it, "the crushing reviews by well-known sociologists convinced me of the necessity of my work because these critics were simply unable to follow the arguments presented."[22] Similarly, though some of Solvay's concepts were at times rather convoluted, they originated from his sincere contemplation of how to attempt to develop a more useful set of societal guidelines, rooted in scientific precepts, that could generally form the basis for the improvement of people's lives. Solvay, like Ostwald, gave scant acknowledgement to others' criticisms of his efforts. After all, the spectacular accomplishment of his adult life had been shaped early on by his willingness to take a theoretical chemical equation and, defying others' failed attempts to commercialize it in practice, make it the basis on which the world's largest chemical company was built. As described in a history of the company, "Flush with his industrial success, which had originally been met with great skepticism, Solvay had developed the idea that his intuitions, even when rejected by the majority, would sooner or later prove right."[23]

Strange as it may seem today, in that era more than a few felt science was equated with the search for laws that, eventually, might be reduced to an overriding principle governing the betterment of society.[24] The more Solvay immersed himself in this train of thought to investigate the existence of a link between physics, chemistry, and biology to uncover a universal rule that governed societal improvement, the more he felt it necessary to establish organizations of research and education on these subjects that could, more specifically, validate his ideas. This culminated with Solvay's founding of a collection of institutes in what became generally referred to as a "Scientific City" in the Leopold Park area of Brussels over the course of ten years as the 19th century gave way to the 20th. The Institute of Physiology was established in 1893, followed by the Institute of Sociology in 1902 and finally a Solvay Business School the following year.[25] These three institutions were built and funded out of Ernest Solvay's own pocket, millions of Belgian francs spent to establish and equip modern facilities and staff them with leading scientific and educational minds from across Belgium, intent on researching to prove out Solvay's hypotheses. As he modestly yet eloquently summed up his motives at that time in a treatise on energy and animal biology, "I have not the good fortune to be a man of science; I have not received a classical education; industrial problems have absorbed my time; but it is true that I have not ceased to pursue a scientific aim, because I love science, and I look to it for the progress of humanity."[26]

Solvay's scientific interests were tremendously diverse, often sending him off in multiple directions of investigation at once. However, the intense focus Solvay had brought to bear in his younger years in solving the industrial problems inherent in his soda production process was noticeably absent when he approached these broader interests. Once again, he became self-educated in numerous scientific disciplines, but he did not become an expert in any of them. Rather, he left it to the professional staff he employed at each institute, an expanding network of forward-thinkers and dedicated researchers, to search out the details and prove his intuitions. Physics, chemistry, and biology had always been Solvay's focus. Now he was intrigued by magnetism, electricity, and atmospheric and low-temperature phenomena as he delved deeper into the wonders of science. He funded experimental work in a private laboratory in Brussels, as well as at a variety of other research locations. Each was supported by Solvay financially, charged with exploring different avenues of study, employing some of the brightest scientific minds in the country, as well as engaging numerous consultants and in some cases building expensive, custom-made experimental equipment for his scientific work. Far from being intimidated by those who might be more educated than himself, Solvay consistently gathered around him those whom he thought could assist in uncovering answers to scientific questions he wished to investigate further.

Solvay had a long-lasting infatuation with gravity and its effects on matter. By the late 1880s, he had completed a paper on the subject, appropriately titled "Gravity," and tried to interest a few of Belgium's leading scientific minds in his ideas on the subject. In his writings, Solvay attempted to relate the law of gravity with his concept of energy in an intuitive approach that he felt could be useful.[27] But the experts criticized this effort due to his lack of formal physics training, which they felt had led Solvay to major deficiencies in proving his points by the selective use of scientific theories that suited his efforts while disregarding others that did not support his arguments.[28]

Solvay was relatively unfazed by his critics, the trademark inner self-confidence in his musings continuing to prod him to intermittently refine his theory over the next twenty years. Just as his soda process was given little chance of industrial success by most in the chemical world, so too Solvay was certain someday that his disregarded scientific speculations on gravity, matter, and energy would eventually be recognized as a significant contribution to science.[29] It was this scientific theory that he wished to expound upon to the assembled brainpower of the Solvay Council, which he had agreed to support when approached about it by Walther Nernst. Once they had heard him out, Solvay was confident that this august gathering would have no recourse but to admit the potential possibilities of his arguments.

So it was agreed. Nernst developed the conference agenda to allow for the sponsor Solvay to open the meeting with some references to Solvay's own keen interest in physics, especially gravity and matter. But Nernst wasn't quite prepared for the detail to which Solvay wanted to explore the topic. As noted by one of Solvay's associates, Édouard Herzen, in a press release that preceded the conference, "Mr. Solvay has been concerned for a long time with the constitution of matter . . . his [thoughts on matter and gravity] reflect his constant preoccupation with the subject since 1887, and even as far back as 1858! . . . this concept of the subject implies a variation in energy, not in a continuous fashion, but by jumps or degrees . . ."[30]

Indeed, Solvay had been considering this topic to the point of postulating the equivalency of energy and matter long before Einstein, as well as their exchanges in discrete "jumps" rather than in a continuous fashion, bearing a strange resemblance to the energy quanta suggested by Planck.[31] Solvay proposed to Nernst that the conference attendees be provided a hefty tome covering Solvay's thoughts on the subject, an in-depth compilation of the theory he had been continuously refining since the 1880s. Nernst was concerned enough about this turn of events to request Max Planck to review Solvay's work beforehand to keep from embarrassing the gracious host of the conference, if not Nernst himself as the force behind the gathering. What followed was a surprisingly good report back from Planck on Solvay's speculations, noting he was impressed that "the author shows that he knows the laws of physics, especially those of planetary motion, in a way that would do credit to a professional theoretical physicist, but also because of his entirely independent and original [approach]."[32]

Solvay proceeded to give the welcoming address to the assembled group, including remarks derived from his preconference mailing. He had previously indicated he would not be a part of the detailed sessions that were to follow due to his not being a scientist. But the stage was set for Solvay's potential involvement in the ensuing discussions when he reviewed his extensive thoughts on gravity, matter, and energy for the attendees, seeking their sage advice, which never came.[33] The audience listened politely, but offered only a muted response that established the tone for Solvay. Henceforth, he would rarely attend the sessions during the week's meetings, even to the point of having his own image pasted into the official group portrait of the conference members after the fact, making his excuses that he had been called away on urgent business at the time of the scheduled picture-taking.

It was only then that Solvay could more clearly envision his true scientific mission in life. He had yearned since his youth to become a scientist, to leave his mark upon the scientific tenets guiding the physical world much as Newton or Maxwell had done. But this was not to be. Instead, for his remaining years he was destined to devote much of his spectacular financial success in industrial chemistry to being, appropriately enough, a catalyst of sorts in the process of scientific progress. In chemistry, a catalyst accelerates the reaction of raw materials brought together to make a product or makes that reaction run more efficiently, in either case being neither consumed nor changed by the reaction but being available to be used again and again in future reactions. Such was to be Solvay's true scientific purpose.

Beginning with the Solvay Conference on Physics in 1911, he would repeatedly bring together the greatest minds of the scientific world to grapple with the most puzzling challenges of physics and chemistry, hosting each of the proceedings in a format unique to scientific conferences of the day, providing a small group of geniuses a focused venue to efficiently debate and chart the course of future exploration. Solvay S.A.'s present-day corporate heritage manager Nicolas Coupain, who has extensively researched and written about Solvay, both the man and the company, states, "Ernest Solvay long believed that scientific progress could be achieved through profound introspection from brilliant-minded individuals."[34] Shortly after the 1911 gathering, he would employ a significant portion of his vast resources to establish Institutes of Physics and Chemistry with initial endowments of one million francs apiece, sponsoring novel research in both subjects. He had never yearned for the spotlight in these or any other ventures, each of which he had always tackled wholeheartedly. His own measure of success had always been in seizing the opportunity that presented itself and determining a way in which he could contribute to bettering society by progressing beyond what had been expected of his own energetic output.

CHAPTER SIX

Einstein's Enigma

In the realm of science, many might identify three specific years in the past six hundred as worthy of the Latin description "annus mirabilis," or "miraculous year." The first two of these are automatically given this designation, acknowledging the extraordinary brilliance exhibited by a specific physicist at a designated point in time. In 1905, Albert Einstein proposed science-altering hypotheses in four papers published on the remarkable behavior of energy and matter at the atomic level. The year 1666 denotes the period of immense creativity and discovery during which the beginnings of basic theories of light, gravity, and planetary motion were formulated by Isaac Newton, as well as his remarkable construction of calculus. Arguably just as significant and thus worthy of this description is the year 1543. At that time, Copernicus gave to the world the heliocentric view of the heavens with the publishing of his treatise "On the Revolutions of the Celestial Spheres," the concepts proposed therein being considered the beginnings of the Scientific Revolution.

In his book on the subject, scientific historian Thomas Kuhn described the impact of the ideas of Copernicus as being far greater than suggesting that the planets revolved around the sun and the calculation of their paths of orbit. As Kuhn states, "Men who believed that their terrestrial home was only a planet circulating blindly about one of an infinity of stars evaluated their place in the cosmic scheme quite differently than had their predecessors who saw the earth as the unique and focal center of God's creation."[1] With one thunderbolt of insight from Copernicus, the accepted status of the natural world would be turned inside out. The human experience would expand to include a limitless universe that was beyond religious faith in heaven and its elements of mystery, establishing more calculable phenomena and birthing modern scientific thought in the process. For his part, Newton's vision of what caused the bodies in the heavens to behave as they did engendered another seismic shift in the human perspective. Einstein's imagination and accompanying calculations, depicting the interactions of the world from an atomic perspective in his papers of 1905, again allowed science to theorize on relationships that had previously been inexplicable.

Copernicus had been ruminating on the relationship of Earth to the sun and the planets for close to forty years before finally publishing his book, but had begun to formulate his ideas while only in his early thirties after completing an extensive university education heavily focused on astronomy. Newton and Einstein were a decade younger at the time of their marvelous years. As has often been the case, the expression of genius determinedly sprung forth from precocious individuals, young minds open to new ideas and delighted in exploration beyond established barriers, creating and verifying visions that often only they could see, at least initially. As opposed to the other two bachelors, the youthful Einstein's miraculous year found him already balancing his scientific investigations with the duties of a husband, of a father to an infant son, and of a job as a patent examiner in Bern, Switzerland. He was not in a position to solely live a life of the mind, but had many competing obligations.

Einstein would eventually become an unparalleled genius in the eyes of the world, but this didn't automatically manifest itself during his childhood in southern Germany. Born into an irreligious Jewish family of modest means, his father a good-natured but ultimately unsuccessful businessman and his mother the stronger-willed of the pair, expectations were not especially great for the young Einstein.[2] He appeared slow to learn and hesitant to speak, initially giving his relatives some concern about his mental capabilities. As Einstein grew, he found he lived in a world in which memorization was the gospel of education and he was a skeptic. He thought in pictures, not words, a process that was at once pleasing to him and almost impossible for his teachers to detect, if they cared to understand his peculiarity at all. Rather, to these instructors his attitude smacked of disinterest, even arrogance, and certainly wasn't suitable for their formal classroom lessons. All this produced a youth who was unconventional, to say the least. He simply did not present the malleable clay his professors wished to mold. However, underneath his careless exterior and ready humor, his mind was fertile ground for ideas, especially scientific ones that fascinated him, and his intelligence slowly began to make itself evident.

As a preschooler, Einstein remembered being shown a small item, something that would instill a yearning to know more about the world around him. It was a compass, a device that, try as he might, he could not determine what inner workings made it function. He recalled, "That this needle behaved in such a determined way did not at all fit into the nature of events which could find a place in the unconscious world of concepts . . . this experience made a deep and lasting impression upon me. Something deeply hidden had to be behind things."[3] As he grew to adulthood, the wonder of this magnetic contraption, directed by forces unseen yet plainly real, was to kindle in Einstein's imagination an eventual burning desire to divine the complex mysteries of the natural universe. He was generally a loner in his efforts, comfortable in this regard, seeking out a select few colleagues to provide sounding boards on his speculations about the curious interactions of the world.

Also at a young age, Einstein's mother introduced music to him, arranging lessons on the violin. After struggling with the regimented practice sessions required to become proficient on an instrument, the process being like a straitjacket for his spirits, Einstein was exposed to Mozart's sonatas and became enchanted with their subtle beauty. As he played the expressive compositions, they provided a serene gateway for his mind to ruminate on many thoughts and questions.[4] Biographer Walter Isaacson captured the infatuation in this manner: "It was not so much an escape as it was a connection: to the harmony of the universe . . . and to other people who felt comfortable bonding with more than just words."[5]

By the time he was entering his teenage years, mathematics was to seize Einstein's attention almost as completely, catalyzed by a geometry book. In it was demonstrated the ability to describe the real world of shapes and forms, circles and spheres, squares and cubes with mathematical equations that were provable and certain, another revelation to his young mind. Although he continued his studies of this subject on into college, Einstein professed that he had forgone delving deeper into mathematics so that he might concentrate in another area that aroused his interest even more, physics. "I had excellent teachers . . . so that I really could have gotten a sound mathematical education. However, I worked most of the time in the physical laboratory, fascinated by the direct contact with experience."[6] Although he was proficient in mathematics, the language of science, in the future Einstein invariably sought assistance from those with a greater mastery of this discipline as he strove to demonstrate his theoretical physics hypotheses with mathematical proofs. He would always say that he was never strong in mathematics.

As he continued his learning, Einstein's unwillingness to adjust to his teachers' Germanic precision and expected standards of behavior represented an affront to their authority that could not be tolerated. He chafed at the bit of rigidity forced upon him in the classroom, eventually dropping out of high school as a result. For a time, he worked in his family's electrical business with surprising success. It seemed his interest in magnetics and electricity, vital to the workings of the machines that the small company produced, made enjoyable the time spent on solving technical problems with most of the devices that needed fixing.

Shortly after, he was determined to go to college. In 1896, after a period of intense self-study and then preparatory school attendance, he gained entrance into a technical teacher's college, the Zurich Polytechnic in Switzerland. There he met someone unlike anyone he had ever encountered, a young woman named Mileva Marić. She was of Slavic origin, from a Serbian family who valued higher education for her, coming to Zurich because as a woman she could not seek university schooling in her own country. Mileva was a bright student, an ardent mathematician and budding scientist who shared her love of these disciplines with Einstein. Like him, she also took pleasure in making music, either on her tamburitza, similar to a mandolin or lute, or on the piano, spending hours in enjoyable concert accompanied by Einstein on his violin as well as other friends with their chosen instruments.[7]

Unlike his first serious relationship, when his attractive looks and jovial demeanor entangled him with a pleasant but superficial young lady, Einstein was drawn not to Mileva's classic beauty, of which she had little, but to her inventive mind and its kinship with his. Although she was not his equal in the theories of physics, Mileva might have been a better experimentalist as well a solid mathematician who could assist Einstein with parts of his study.[8] He found that he could share his scientific ideas with her, and she not only understood them but could discuss them with thoughtfulness and practicality. They took the same math and physics courses, studied together, and shared their thoughts and views on these subjects as well as life.

Letters exchanged between school terms as students at the Polytechnic show the conversations of the pair transitioning over time from details of homelife and classwork to more complex ideas and playful intimacies. In the early months of their relationship, the breadth of Mileva's thoughts expressed themselves in a note where she mentions the concept of infinity. While she was attending classes for a semester in Heidelberg, Germany, she wrote to Einstein of a foggy view of a forest she was visiting outside of the city, "all I see is a certain something, desolate and grey as infinity . . . Man is very capable of imagining infinite happiness, and he should be able to grasp the infinity of space. I think that should be much easier."[9] Over the next few years, as the century turned, in letters to each other she had become Doxerl, or "Dollie," to his self-stylized Johonzel, "Johnnie." An Einstein letter to Mileva from the summer of 1900 began with an amusing attempt at a poem, the first four lines of which were

> O my! the Johonzel,
> He is totally crazy.
> He was thinking about his Doxerl
> And he squeezed his pillow.[10]

Yes, Einstein was in love. This, despite warnings from his family, who scoffed at him for making a mistake. From their superficial vantage point, he was taking up with a woman who was a few years older than himself, who walked with a limp from a congenital hip condition, and seemed to have a much more serious disposition than he. Although his parents resolutely avoided practicing their own Jewish religion in an effort to assimilate, they commented on Mileva as not being Jewish, worried that a mixed marriage might make things even more difficult for them in society. They could only see future pitfalls in committing to this relationship, which served to heighten Einstein's will to do what he pleased. For a time he listened to the complaints being registered by his parents and then, much as any strong-willed son might respond when pushed in a direction not to his liking, went his own way.

When they both took their final exams, Einstein was awarded his technical teaching diploma but Mileva was not because her test scores did not measure up. It was a disappointment to her, but they were excited for Einstein to now take his first professional step. He began to seek steady employment as a science instructor or research assistant, in his search being frequently apart from Mileva. But he had no success in obtaining a position, not even a glimmer of a positive response from numerous inquiries. Einstein's teachers had experienced his famously irregular habits while a student. This, along with his perceived arrogant demeanor, resulted in a lack of professorial connections who would provide glowing recommendations of his past efforts in their classes. Suspected anti-Semitism added to the obstacles. So he lived by accepting tutoring jobs when available, augmented by his relatives' beneficence when offered.

During this time, the two sent numerous letters back and forth discussing the myriad thoughts swimming in Einstein's brain about the universe and how it was ordered. Over the next few years, Einstein would correspond with Mileva about his concepts related to a number of physical phenomena, some of which, from his letters' wording, appeared to be focused on project work that might have been shared by the two of them. In one letter he wrote, "we shall seek to get empirical material on the subject [of capillarity] . . . If a law of nature emerges from this, we will send it to Weidemann's *Annalen* [German physics journal]."[11] In later correspondence, he asked Mileva to see if she could research and develop for him some information about thermal properties of glass to support some new ideas Einstein had concerning energy.[12]

His communication to her a few days later started with a typical review of his plans to gain an academic position, possibly with the help of a friend whose uncle was a mathematics professor. As he laid out his strategy, Einstein veered off topic with a quick aside to profess his undying love for Mileva. What followed was, in hindsight, one of the most unusual, most referenced and most speculated-upon exchanges between the two: "How happy and proud I will be when the two of us together will have brought our work on the relative motion to a victorious conclusion!"[13]

This seemingly tangential statement in the letter was one that has perplexed historians and physicists for decades, beginning when the English translation of the first volume of Einstein's letters was made available to the public in the late 1980s. *The two of us together will have brought our work on relative motion to a victorious conclusion!* The implication appears to be that both Einstein and Mileva had worked in concert on a theory of relative motion, which could be construed to mean the special theory of relativity, on which Einstein alone would publish one of his most famous papers in 1905. Physicist Pauline Gagnon is one of a number of believers in Mileva's contributions to Einstein's work, making the case for many others, in this reference to relative motion, "nobody made it clearer than Albert Einstein himself that they collaborated on special relativity when he wrote to Mileva [with this statement on relative motion]."[14] Since the publishing of this and other letters between the two referencing potentially joint scientific efforts, the argument has raged back and forth between each faction attempting to establish Mileva in her correct place as a collaborator, or not, with Einstein on his famous theories that were published in 1905.

The temptation to interweave the scientific construction of the special theory of relativity as the joint mental knitting of these two young scientists has been a strong one. When first discovered by researchers, the letter's contents combined with a number of other letters where Einstein spoke to Mileva of "our work." This added to oral histories handed down by some of her family members and close friends as well as a few Eastern European scientists to claim an inventive presence for Mileva in the development of Einstein's groundbreaking theories.[15] Conversely, many just as emphatically asserted that these arguments represented weak grounds for Mileva's claimed coauthorship of greatness, especially since she apparently never offered evidence herself of potential formal collaboration. Much in this regard can be summed up in a few thoughts from history professor Alberto Martinez in an article that specifically countered the argument about Einstein's quote in his letter to her concerning their work on relative motion. "Non-specialists might conclude that this statement refers to relativity theory. But it does not. At the time, Einstein believed in the existence of the ether [a much-debated medium that was generally accepted to transport electromagnetic waves]. He wanted to devise experiments to test its relative motion—a puzzle that drew the attention of many physicists."[16] Needless to say, to this day there has surfaced no definitive answer to the enigma of Mileva's role, if any beyond constructive discussion, in formulating Einstein's theories.

Nonetheless, the precedent for a husband and wife pair of scientists, each sharing their most inventive ideas and supporting the others' speculative thoughts, though outside the mainstream and generally far from the realm of possibility, was not unheard-of. While in college, Einstein and Mileva had potentially followed the discovery of radioactivity by none other than Marie and Pierre Curie. Perhaps she had hopes that this would be the blueprint that their lives could follow, her scientific capabilities complementing Einstein's physics ingenuity, working together to unravel the deepest secrets of the universe, the Swiss version of the French Curies. To be sure, just as her fellow Slav Marie Skłodowska Curie had done before her, Mileva had seized her opportunities to study abroad and advance within the academic world. She had fallen just short of becoming university-degreed, but in the bargain Mileva had expanded her education and formed a connection with a brilliant individual who appeared to want to share his scientific thoughts as well as his life with her. But was he willing to support her own determined efforts to achieve a life of science?

The bond between the two lovers continued to deepen. Then, after a romantic holiday spent at Lake Como, Italy, Einstein fathered an illegitimate daughter by Mileva in early 1902. Heartbreakingly, the baby girl, named Lieserl, was lost in undocumented history as perhaps put up for adoption or dying as an infant. There was no record of the actual events, just a few vague references in letters and Marić family lore. To his credit, Einstein did not flee the scene; he married Mileva instead, eventually having two sons with her as they seemed to settle into a life of domestic contentment, if not bliss. For all of her effort, Mileva never received her scientific degree, failing in a second attempt at her exams and moving on to a life with her "Johonzel." However, things did not evolve into the relationship she had envisioned for herself with her university sweetheart.

Through a friend's father Einstein finally found a permanent job, taking an entry-level position as a patent examiner third class within the Swiss government bureaucracy after marrying Mileva in 1903. He had treated her as an intellectual equal before their marriage, but now Mileva was relegated to what Einstein deemed her proper place as a wife, cook, housemaid, and soon-to-be proud mother of a baby boy the next year. He began to rely on Mileva less as an intellectual foil for his ideas or for her mathematics skills that might assist in developing some of his scientific proofs, turning to a small group of male friends to debate and design physics theory. She bemoaned the changes but was powerless to do more than mildly protest. And society's norms gave Einstein every reason to treat her in this manner. That he chose this path for his wife spoke volumes of his lack of desire to promote any of her continued striving toward scientific achievement.

Mileva gave birth to a son in 1904, with finances tight and neither of their parents nearby to potentially provide childcare so she might have some freedom to pursue her own career interests. In contrast, Pierre Curie's father had so ably stepped in to care for the couple's first daughter, allowing Marie to continue on her scientific journey of discovery. Thus, the final nail was put in Mileva's professional coffin, making it almost impossible for her to take a scientific breath when the lid was hammered into place. She did her best to carry on in support of her new marriage and family, but no doubt yearned for more.

Einstein scarcely noticed, his existence being transformed rather quickly from bohemian to bourgeois. He jumped into his new position and was a quick study at the patent office. He was immediately intrigued by the numerous applications that crossed his desk concerning various devices. Many were electromagnetic in nature, having been developed by eager minds awaiting patent approvals so that the inventors might cash in on their ingenuity. From his prior efforts with his family's electrical product manufacturing business and his scientific education, he could easily understand and determine the attributes and potential defects of each application. He busily processed his daily workload in a few hours so that he could use the balance of his time to investigate his own scientific ideas.[17]

And what concepts he wished to mentally explore! One recurring vision from years earlier had him traveling at the speed of light, astride a light beam that was moving in parallel to another beam that he was observing while attempting to overtake it. From where he sat, Einstein's view of the other beam might appear to him to convey that both beams were stationary, yet he knew both were moving at the speed of light. Another episode placed him standing in an elevator that was accelerating as it moved higher and higher in the emptiness of weightless, deep space. The rocket moved faster as it sped upward, while he was monitoring the floor, which seemed to be pressing against his feet. It appeared he was experiencing gravity, yet in outer space he knew this could not be. How to make sense of these contradictions? These and many other thoughts presented many questions. More important, they could be translated by Einstein into the forces of nature to which all of us are subject. The brilliance of his mind was not only in the imagining, the constructing of these "thought experiments," as they were called. It was in the understanding of what was taking place, and what it meant for human existence. Einstein focused on what to most was imponderable, either because the things he was able to envision were too complex for anyone but a genius to conjure or because the curiosities he examined were too mundane for it to occur to people to spend time contemplating them. From these musing were drawn the seeds of many of his theories.

By 1905, Einstein was in the midst of what was the most domestic period of his adult life, coming home to dutifully push his baby boy in a pram after finishing a day at the patent office. At the same time, he was in various stages of completing his doctoral thesis as well as four papers for submission to an esteemed German scientific journal, *Annalen der Physik*, each of which would challenge conventional thought concerning their topics of investigation. They delved into phenomena that had intrigued scientists for a number of years, in some cases questions that were beginning to fray the fabric of classical physics.

In the spring, Einstein sent a letter to a friend, proposing what now might be seen as the most lopsided trade of scientific concepts (from a value perspective) ever suggested. He requested his friend's doctoral dissertation and in return offered to send copies of four papers, his own dissertation as well as three other articles being published that year in *Annalen*. One of these three dealt with energy quanta in relation to the photoelectric effect, another discussed the forces behind the random motion of small molecules in a liquid, resulting in effects on discrete particles suspended in the liquid, while the last was a rough draft of a paper that hypothesized special relativity.[18]

The only paper of Einstein's from 1905 that was missing from this proposed swap was one that he would author later in the year that equated energy with mass as an extension of his work on special relativity, leading to the most famous equation of all time, $E=mc^2$ (energy equals mass times the speed of light squared). His calculations and results on molecular effects on small particles suspended in a liquid laid the groundwork for the definitive establishment of the physical existence of atoms and molecules. The paper on the photoelectric effect provided the basis for his being awarded the 1921 Nobel Prize in physics. Taken together, these various works combined to form the most earthshaking hypotheses for physics since Newton's efforts during his miracle year over two centuries earlier.

At that time, Newton had been an earnest but unremarkable student pursuing a master's degree when his expression of genius burst forth as he wandered the lonely grounds of his Woolsthorpe manor. Similarly, Einstein in 1905 was relatively secluded in the Swiss town of Bern, content with his new family, practicing his patent-examining craft, far from any major scientific center of learning. Each had spent significant time in self-study of scientific theory as presented in voluminous texts. Neither had attained a professorial post, let alone was recognized as brilliant. Yet both, in their own ways, were able to perceive something of the universe that others could not. Whereas Newton's views were rooted in what he could physically observe, an apple dropping from a tree, Einstein's thoughts sprung from the pictures he generated in his mind's eye, careening through space perched on a shimmering beam of light. As only genius can, both focused intently on their images to discover hidden meaning and interconnectedness that led to the development of amazing theories about the physical universe.

In May 1905, Einstein submitted a paper to *Annalen* focused on light and its energetic composition. Hundreds of years earlier, the nature of light had been originally postulated by Newton, though not proven, as being composed of minuscule particles. However, over the course of time it had been generally accepted that light was made up of continuous waves rather than individual particles due to results of various experiments in the scientific community. By the mid-1800s, the work of physicists Michael Faraday and then James Clerk Maxwell emphasized this wave nature of electromagnetic radiation, light included. Even so, some of the experimental work of others did not fit neatly into the wave theory of light, and as more tests were conducted, this conundrum became more significant. A particular instance that intrigued many physicists concerned something as simple as the shining of light on a metal surface.

Experimentation concerning electrons had become more commonplace since their discovery by English physicist J. J. Thomson at the end of the 19th century. In the illumination of some metals with light, electrons could be detected as being emitted from the surface of the metal. This was termed the photoelectric effect, which the wave theory of light could not adequately explain. Classic physics thought could not resolve why only light frequencies above a certain level caused electrons to be emitted, while light of lesser frequencies had no effect at all in causing electron emission. Nor was the emission of electrons from the metal's surface specifically dependent on the brightness of the light. Physicists grounded in two hundred years of the accepted wave theory of light, most recently supported by Faraday and Maxwell's experimentation, were puzzled.

Einstein, taking a step back from the entrenched dogma of the period, speculated on the possibility that the nature of light might be different from its accepted wavelike structure. He had been following the work that German physicist Max Planck had announced to the world five years earlier concerning the potential for light to be composed at times of either waves or particles, discrete energy packets he called quanta.[19] Einstein now postulated that light was not only wavelike in nature, whose energy would be spread over the whole of the wave, but it could be discontinuous as well, composed of discrete particles, each containing a specific amount of energy. As Einstein put it, "in the propagation of a light ray emitted from a point source, the energy [from it] . . . consists of a finite number of energy quanta . . ."[20] This would explain the situation nicely.

According to Einstein, energy from the light particles, or quanta, was absorbed by the electrons of the metal. If the frequency of the light particles provided enough energy to the metal electrons, they could escape the bonds holding them to the metal, retaining any excess energy from the light particle beyond this escape energy. If the frequency of the light particles possessed insufficient energy to pass on to the electrons they encountered in the metal, the electrons simply would not absorb enough energy from the interaction with the light quanta to escape and be emitted from the metal.[21] Understanding the photoelectric effect in this manner, so radical at the time yet so elegant in solving the issue, eventually became the basis on which photoelectric systems today generate power for devices from light and motion detectors for automatic door openers to the operation of solar cells.

This quantum hypothesis, initially proposed by Planck as a theoretical mathematical solution to a thermodynamic issue on which he had been wrestling, was now shown by Einstein to be an explanation to a real-life physics mystery, the type of issue he was eager to tackle through unconventional thought. However, being the theoretical physicist that he was, Einstein himself did not provide the experimental evidence to support his contention. Not uncharacteristically, in his paper he states rather plainly that "I wish to present the train of thought and cite the facts that led me onto this path, in the hope that the approach to be presented will prove of use to some researchers in their investigations."[22]

Indeed, over the next few years a number of experiments were performed by others that led to early validation for his quantum thoughts. However, it took over ten years, until American physicist Robert Millikan's experimentation in 1916, for enough data to be generated to finalize Einstein's pronouncement as a valid theory.[23] But Einstein's revelations in this paper, combined with Planck's initial speculations on quantum energy, were enough to generate a substantial crack in the bedrock structure of heretofore accepted theories of the physical world. This fissure became the crisis in physics that generated the major discussion topics of the First Solvay Conference on Physics in 1911.

In his second paper, Einstein tackled another peculiar issue that had been a scientific conundrum for quite some time but was not formally examined until the late 1820s. At that time, English botanist Robert Brown carefully documented his observations that, under a microscope, tiny bits of plant pollen suspended in a still liquid moved in erratic, seemingly random motion. This was given the name Brownian motion, an odd phenomenon with no concrete explanation, though many were proposed. Einstein's thoughts on the issue led him to employ kinetic theories of gases and diffusion theories of matter to the problem. He hypothesized that the liquid was composed of particles much smaller than the pollen bits, not visible even under the microscope, which were bumping into the pollen particles and causing them to randomly move in different directions. These invisible particles were atoms or molecules, comparatively huge in number and insistently bouncing off the pollen bits and causing them to move in an irregular manner.[24]

The idea of the existence of atoms and molecules had been previously proposed by numerous leading researchers, but a significant contingent of the scientific community steadfastly refused to believe in their existence, in no small part because these minuscule particles were not readily detectable by sight nor able to be measured. Some took to employing the idea of atoms and molecules in their theoretical work, but still viewed the concept as only hypothetical. This led to Einstein's approach being thought by some to be controversial. As physicist David Cassidy noted, "Einstein predicted that the random motion of molecules in a liquid impacting on larger suspended particles would result in irregular random motions of the particles . . . from this motion Einstein accurately determined the dimensions of the hypothetical molecules."[25] This mathematical determination of atomic sizes would be crucial in substantiating Einstein's proposed theory. As with his article on the photoelectric effect and energy quanta, Einstein ended his paper on Brownian motion with a challenge to the experimental physicist to provide detailed experimentation and verifiable data: "Let us hope that a researcher will soon succeed in solving the problem presented here . . ."[26]

By 1908, the French physicist Jean Perrin did just that. He performed a precise series of experiments that generated substantial information in support of Einstein's hypothesis on the cause of Brownian motion as being atomic collisions, showing Einstein to be correct in his calculations and predictions.[27] This was essential to convincingly prove the existence of atoms to most doubters, eventually winning for Perrin the Nobel Prize in physics in 1926 for his efforts.

The last two papers Einstein submitted to the *Annalen* in 1905 were based on efforts exploring the essence of the fantastical light beam visions that he had first pictured many years before. This image had captivated Einstein since that time, and he had been consistently pondering this depiction and working out an explanation for its significance over the years since. With the principle he developed, Einstein was to show that the relationship of space and time was not what it had appeared in the past, when each was thought to be absolute and finite for everyone and having no interrelationship. He was to clearly state that the connection between the two depended on the frame of reference of the observers of any specific event.

Einstein began by comparing the vision of two bodies moving relative to each other, and took as a starting point only the case in which the speed of each of these two bodies was constant and was moving in a straight line, meaning that neither one would be accelerating or decelerating, nor would they be moving in a curved direction. Hence, the modern designation of *special* theory of relativity for this work was a special case with no acceleration or deceleration involved (those conditions would be examined in the future by Einstein as he developed his *general* theory of relativity). He then took as a given a reasonable assumption that the laws of physics were the same for the two bodies in his vision. Lastly, he assumed that the speed of light was finite at 186,000 miles per second, completely independent from the motion of its source. Though counterintuitive, it had been postulated decades before and was proven to be so time and again in multiple experiments since.

With these three principles, one could examine various theoretical situations, what would be termed "thought experiments," to investigate the interrelationships of the two bodies to events affecting both. This would allow an examination of the experiences of each body *relative* to the other. One classic thought experiment proposed that the two bodies were two people, one stationary and the other in a nearby train moving in a straight line at a constant speed. It could be shown that if lightning were to strike two trees simultaneously, the trees being equidistant from the stationary observer and with one tree toward where the train was moving and one in back of where the train had passed, the two people would perceive the event differently. To the stationary person, it would appear that the lightning strike was simultaneously hitting both trees, but not to the person in the moving train who was passing the stationary person exactly at the time of the lightning strikes. This was because of the different frame of reference of each person, the person in the moving train seeing the lightning strike of the tree toward which the train was moving a split second before the lightning strike of the tree away from which it was traveling due to the velocity and direction of the moving train. Thus, time was not absolute (that is, time was not the same for each person) in that one saw the simultaneous lightning strikes of the two trees while the other saw the forward tree being hit by lightning at a time just before the other was struck. Time actually was dependent on the state of motion of the observer. Ultimately, special relativity related two frames of reference as they viewed a specific event, stretching time and space in the process.[28] Simply put, time had been affected differently for each of the individuals in this thought experiment. It had actually expanded for one by the smallest fraction of a second due to that person's relative frame of reference vis-à-vis the event both had witnessed.

Further, in a follow-up paper Einstein explained that a consequence of special relativity ultimately manifested itself in the equivalence of mass and energy with the equation $E=mc^2$ (energy equals mass times the square of the velocity of light). During the summer of 1905, Einstein wrote another letter to the same friend to whom he had proposed the swap of papers. This time, he noted his idea about the relationship of mass to energy, ending in a typically whimsical manner: "The relativity principle . . . requires that the mass be a direct measure of the energy contained in a body . . . the consideration is amusing and seductive; but for all I know, God Almighty might be laughing at the whole matter and might have been leading me around by the nose."[29]

As it turned out, the serious consequences of this theoretical relationship are today well known worldwide. Einstein's equation showing the equivalence of mass and energy indicated that even a small amount of mass could harbor a huge amount of energy. This would be demonstrated in real life by nuclear reactions that can, under controlled circumstances, generate energy for the world's use via nuclear power plants as well as facilitate the tremendous destructive power of nuclear bombs. Eventually, Einstein was to recognize both possibilities, especially the weapons issue. During his later years he would somewhat idealistically urge the need to create an "empowered world government" to regulate the use of what he saw as the inevitability of nuclear weapons.[30] As in another age, when Alfred Nobel invented the terrifyingly convenient material called dynamite, the power of scientific invention was shown again to be a double-edged sword. At that time, Nobel's remedy for his brainchild's potential harm to humankind was to set aside a portion of his fortune as a prize, an annual incentive to support peace in the world. Never a wealthy man, Einstein could only offer his lifelong dedication to worldwide pacifism to combat the coming nuclear arms race.

These final two published writings of significance from Einstein's miraculous year had explored the special theory of relativity, which would be shown to be a subset of the general theory of relativity that he would develop over the next ten years. Relativity, in both its variations, along with the energy quanta proposal that became a precursor to quantum theory, would be seen as Einstein's signature scientific contributions to the world. From the 1920s onward, it would be impossible to imagine the scientific landscape without the mention of one of these original insights that had emanated from the mind of Albert Einstein.

Certainly, Einstein's light quanta work was immediately recognized by Max Planck, an editor of the *Annalen*, and a few others as being extraordinary in nature. This was especially so for Planck, who had surmised the energy quanta five years before in his theoretical problem-solving efforts concerning black body radiation and had been somewhat ambivalent about its actual existence ever since. By 1911, he and Walther Nernst were to decide, with the support of Belgian industrialist Ernest Solvay, that it was time for an elite group of scientists to debate the significance of the energy quanta as part of a greater review of the upheavals fracturing the classical tenets of physics. The time to convene a special physics conference was at hand.

CHAPTER SEVEN

The First Solvay Conference: Assembly of Genius

Within a few months of each other in the second half of 1911, two of the most famous women in Paris vanished. One had fascinated the world with her amazing discoveries while the other had captivated the public with just a smile. Both were not originally from France but each had come to Paris and made it their home. To many, one represented everything that was admirable in women, the other most everything that was problematic—at least according to the sensationalistic French press as it trumpeted the disappearance of one, then the other.

As the summer unfolded that year, the weather was unbearably hot across Europe. But nowhere was it more scorching than in Paris. From early July through mid-September, the temperatures soared. At times they climbed to 40 degrees Celsius (104 degrees Fahrenheit) or more during the day and, for a two-week stretch in August, rarely dropped below 30 degrees Celsius (86 degrees Fahrenheit) at night, causing over 40,000 deaths from the heat, mostly babies and the elderly.[1] Yet, on Sunday evening, August 20, a thief hid himself away in what must have been a suffocatingly warm supply closet in the Louvre Museum in Paris and sweated out the hours. On Monday morning he escaped his self-made oven wearing a maintenance man's smock and stole the *Mona Lisa* while the museum was closed to the public for its weekly cleaning.[2]

The portrait, though iconic, had not yet become the most well-known painting in the world. Yet its theft and the subsequent global newspaper reporting about it, complete with photographs, quickly propelled it to those heights. It was one of Leonardo da Vinci's most popular works and much appreciated as it hung in a heavy wooden frame behind a newly installed protective glass enclosure on a wall of the Salon Carré in the Louvre. Its subject's subtle beauty, grace, and hint of a smile entranced countless admirers. Da Vinci had labored for many years over the image of what was thought to be the young third wife of a successful Florentine merchant, Francesco del Giocondo. Finally, near the end of his life, after continued efforts on the portrait, da Vinci declared it still unfinished. His last patron, King Francis I of France, greatly admired da Vinci's work and acquired the painting. Since that time, it had been a part of French culture, for a period even gracing the bedroom suites of the Emperor Napoleon before finding its current resting place in the Louvre during the early 19th century.

As only the French can put it, the young woman's visage had a certain "je ne sais quoi," an indescribable quality that invariably enthralled the crowds. Her look of supreme calmness and the faint, contented smile combined to transmit a feeling as well as an image. Many sensed joy or happiness in observing the lady, hence the alternative French designation for the painting, *La Joconde*, the jovial one, a play on the name of the Italian Giocondo whose wife was believed to have posed for the picture. The gentle, searching eyes transfixed the viewer, who sensed

Mona Lisa was gazing solely at him or her. Da Vinci himself believed that the "eyes [are] the mirror of the soul."[3] He had captured in her eyes, along with her accompanying smile, something stirring and unique, some say a quality that spoke to the very essence of virtue. The serenity of the *Mona Lisa* evoked previous depictions of the Virgin Mary, the image of morality to which all women should aspire, at least as thought by many in polite society. Taken in its totality, the painting transmitted an aura of beauty and purity of subject, truly a woman worthy of being put on a pedestal as the ancient Greeks and Romans had depicted their female gods in marble sculpture.

All of Paris, and much of the world, was stunned by the audacity of the disappearance. French newspapers ran headlines expressing the shock, exclaiming, "UNIMAGINABLE!" "INEXPLICABLE!" "INCREDIBLE!"[4] Who would dare to make this brazen attempt to steal away from France one of its most idolized women? As R. A. Scotti's description in her intriguing book on the subject explained it, "When Mona Lisa slipped out of her frames, she seemed to change from a missing masterpiece to a missing person . . . the public felt her loss as emotionally as an abduction or a kidnapping."[5] Scores of investigators were put on the case to follow up on leads from every corner of the country and beyond, none of which proved helpful. Some wondered if an unscrupulous American millionaire was behind the robbery, perhaps J. P. Morgan himself or one of his cronies, flush with robber baron cash and intent on using it to help themselves to a piece of European beauty and refinement. Finally, after over two years of fruitless efforts, the beloved lady was recovered when the culprit tried to sell her to an art dealer in Italy.

Later that fall, the disappearance of the other Parisian woman of notoriety was that of Marie Curie. Granted, she was not abducted and secreted away in the early morning hours, but her absence from Paris was noted nonetheless in stunning fashion a few months later. On Saturday, November 4, the front page of one of the largest daily newspapers in Paris, *Le Journal*, carried the startling headline: A STORY OF LOVE: MADAME CURIE AND PROFESSOR LANGEVIN.[6] A reporter had written a gossipy article on the two scientists, who had known each other for years dating back to when Paul Langevin was Pierre Curie's protégé at EPCI. He spent most of the piece conveying what he had gleaned from an interview with Langevin's mother-in-law over a café au lait. The author highlighted the story by referencing potential love letters written between the two. Then he ended by posing an open question to the pair of supposed lovers as to the veracity of it all, asking that they refute the situation if they could and closing with a statement that "Mme Curie can't be found and no one knows where to find M. Langevin."[7] Indeed, there was no trace of either of the two physicists in Paris that autumn day. But their whereabouts were not a mystery to their colleagues, friends, or families. All had known that they were in Brussels for the week, attending the First Solvay Conference on Physics.

The riddle of Marie Curie's disappearance was not a mystery at all, purported to be so only to sell newspapers. In France, this was the golden age of sensational journalism, so the more outrageous the headline, the more papers were sold. What better story to peddle than catching two famous Parisians in a clandestine love affair, running off somewhere, perhaps eloping? Although Curie was to adamantly state that her private life was nobody's business, the month of November saw the gradual erosion of this knee-jerk response of trying to protect herself and family. The conservative press felt confident in taking direct aim at a foreign-born, independent woman who had succeeded in a man's scientific role, one whose goals could be twisted into a story of homewrecking and destruction of native French values. Even a woman whose not-so-distant triumphs in the discovery of radioactivity had received resounding applause from the French citizenry—who at the time, of course, were very quick to claim her for their own. For much of the world Marie Curie represented the peak of scientific achievement by being awarded the coveted Nobel Prize. A woman whose presence was insisted upon by Ernest Solvay as he scanned Walther Nernst's carefully selected attendee list for their planned physics congress and found it missing.

The French press had only months earlier pointed to the vanishing of the *Mona Lisa*, with its idealized femininity, as a tremendous loss to French culture. Now the conservative newspapers in Paris trained their sights on destroying the reputation of the missing Marie Curie as a symbol of women who did not know that beauty, modesty, and deference to masculinity were the standards by which France measured her respect for women. Curie had been made to run this same gauntlet only ten months before when she had suffered through a gruesome character assassination at the hands of a xenophobic, misogynistic press and was denied a place in the Academy of Sciences. Then, as now, tinges of the Dreyfus scandal hung over the accusations cast at Curie. This in large part reflected the country's decades-long internal recriminations with its own narrow-minded, anti-Semitic, nationalistic pride that allowed the innocent French captain Alfred Dreyfus, who was Jewish, to be wrongly convicted of treason as a supposed undercover spy for Germany against France. Now, the charge of improper affection with a married man was added to Curie's crimes of not only being an ambitious woman and a mistakenly suspected Jew, but of not being truly French at all in the eyes of a nativist press.

The week had begun so much more auspiciously for Marie Curie and Paul Langevin as they joined three other noted French scientists, Henri Poincaré, Jean Perrin, and Marcel Brillouin, at the First Solvay Conference on Physics on Monday, October 30, at the Hotel Metropole in Brussels. The French contingent was all from Paris and mostly part of the same tightknit scientific community there. Of the delegation, each either taught at the Sorbonne or across the street at the College of France, the country's premier research institution. A number of this group even took summer holidays together with their families on the French coast in a part of Brittany that teemed with Parisian scientists and professors. Perrin had been a next-door neighbor of Curie's when both resided on the Boulevard Kellermann in a Paris suburb, a scientific peer who had been a classmate of Marie's at the Sorbonne and would go on to win a Nobel Prize in physics. Poincaré had been one of Curie's professors when she first attended the university, a revered mathematician of immense stature in the academic world. Brillouin was a steady if unspectacular scientist, well thought of as a teacher and friends with Langevin and his family. And Paul Langevin himself was a brilliant physicist who had studied under Marie Curie's husband Pierre, had taken over his teaching position at EPCI when Pierre was elevated to a full professorship at the Sorbonne, and through insightful research was giving every indication of perhaps being the next great French scientific mind. Shortly after Albert Einstein published his theory on the direct relationship between mass and energy in 1905, Langevin, working independently, had arrived at substantially the same conclusion, missing out on being first to publish by a matter of months. He became an ardent supporter of Einstein's "special theory of relativity," as it would eventually be known, enthusiastically teaching it to his students as the latest in the wonders of physics.[8]

By the time of the Solvay Conference, Marie had known Langevin for more than a decade, their families casually interacting and the two becoming closer a few years after her husband's death while Langevin's marriage to an increasingly abusive spouse continued to unravel. Their friendship had evolved to become intimate over the past few years, two brilliant but lonely people searching for a more meaningful human connection. They had done their best to keep this relationship quiet but, as is often the case within a small group of associates, the interconnections of the French scientific community quickly made more public than anyone desired certain personal details passed on to others in confidence. Hence, the Curie-Langevin affair was known, to varying degrees, to these three Parisian attendees accompanying them as all arrived at the Hotel Metropole. The question was whether or not its deteriorating veil of secrecy would become more fully undone.

The German attendees to the congress were an even more illustrious group. Besides Nobel Prize laureate Max Planck and future winner Walther Nernst, fellow Berlin physicist Heinrich Rubens came along, as did another recently crowned physics Nobel Prize winner, Wilhelm Wien. Arnold Sommerfeld joined from the University of Munich, a superb theoretical physicist whose Nobel Prize fate was to be the eternal bridesmaid, being nominated a record-setting eighty-four times but never capturing the honor. Rounding out the participants from Germany was physicist Emil Warburg, who, although never being awarded a Nobel Prize himself, fathered a future Nobel Prize winner in medicine, Otto Warburg.

Two participants each came from England (including Nobel Prize winner Ernest Rutherford), Austria (one of them being future Nobel winner Albert Einstein, who as a professor from Prague was required to obtain citizenship in Austria-Hungary), and Holland (with the facilitator of the conference, Nobel laureate Hendrik Lorentz, as well as future Nobel winner Heike Kamerlingh Onnes). One had also come from Denmark. Altogether, nine of the eighteen scientific geniuses who had accepted Solvay's invitation to assemble for a week of discussions over the future of how physics should view the natural world would become part of Nobel Prize–winning history. Including Solvay and two assistants, as well as three other scientists who were charged with administrative and secretarial duties for the gathering (Belgian Robert Goldschmidt, Frenchman Maurice de Broglie, and Frederick Lindemann from England), twenty-four individuals were invited to an inaugural dinner reception on the eve of the conference.

They were anxious to hear each other out, contest the issues, and try to make some sense of it all.

Most of the guests had responded quite enthusiastically to Solvay's invitation. Nernst summarized the positive comments from some of the individuals in a letter to Solvay before the conference. He noted, "Rubens told me that he found your kind invitation just grand. Rutherford declares that the idea is excellent. Kamerlingh Onnes has especially asked me to inform you of his appreciation for the imaginative new direction which your beautiful idea of the conference would give rise to. And Einstein is just delighted with the whole thing."[9] Marcel Brillouin was even more specific in his letter back to Solvay, commenting on the heart of the plan to have only a select few scientific thought leaders present to discuss the crisis facing physics. "The initiative which you have taken appears to me of the highest interest. It is in a meeting of limited [number of participants] such as the one you envisage that it is possible to exchange profitably one's views about new ideas, which are far from being clear."[10]

The very concept of this conference, its focus on a defined, narrow set of perplexing issues in physics, to be considered by the premier minds in physics from across Europe, was quite appealing to those involved. They were accustomed to infrequently attending more traditional gatherings that were invariably open to any scientists in a specific discipline to submit papers that would be presented to a large audience with no chance of formal discussion or debate. The Solvay Conference was so novel that it was immensely attractive if only in its creative approach. Hendrik Lorentz, the universally respected Dutch physicist who had graciously accepted the daunting task of facilitating all of the sessions for this gathering, wrote a letter to Solvay during the summer before the scheduled meeting to express his own thoughts on the unique conference forum. It read in part, "Allow me to tell you also that I greatly appreciate your generous efforts for the progress of science, and your project has my complete sympathy. In fact, this meeting of a limited number of physicists, all of whom are seriously occupied with the important questions on the programme, would serve without doubt greatly to elucidate and to state the difficulties precisely, as well as to prepare the way for their solution."[11]

Lorentz was becoming accustomed to being called on for his leadership abilities by the scientific community for what was a newly developing trend of international conferences that had started to take place as the new century began. In a compendium of essays honoring Lorentz that was assembled by his daughter after his death, she noted firsthand her father's characteristics that placed him so much in demand for this role: "All scientists were amazed at the all-encompassing breadth of his understanding and at the ease with which he was able to explain subtle points . . . Add to this his well-balanced character, his tact in meeting people, [his ability to] resolve differences sure to arise during large conferences, and last but not least his perfect knowledge of modern languages . . ."[12] In short, he was the ideal choice to be the leader for a multilingual gathering of an international group of brilliant scientists who had differing opinions often accompanied by inflated egos.

One such scientist was Walther Nernst. He was thrilled with the roundly positive reaction to Solvay's invitation, alleviating any concerns that had been expressed by Planck in their previous correspondence wherein Planck had shown serious reservations about potential interest from others in such an endeavor. For, as much as this conference bore the name of Solvay and was financially underwritten by him, most if not all knew it as Nernst's idea. Similarly, many of the guests could suspect his desire for this conference to be a successful undertaking to help lay the groundwork for his push for Nobel Prize stardom by connecting Nernst's latest work with that of Planck and Einstein on quantum theory. Nernst had a long-standing professional rivalry with the distinguished Swedish chemist Svante Arrhenius, a Nobel laureate himself as well as one of the powers behind the Nobel Prize Committees for Physics and Chemistry, who appeared to often stand in the way of Nernst receiving this coveted honor. By connecting his scientific investigations to those of the quantum pioneers, Nernst felt he could expand his professional credentials, mainly based in experimental chemistry, to include being part of the leading edge of physics. In so doing, he could continue to build a case for his deserving a Nobel Prize in spite of any obstacles Arrhenius might place in his path within the committee's deliberations.[13]

As everyone agreed, none more so than himself, Nernst was certainly a scientific genius. Accompanying this self-awareness was more than a touch of arrogance that at times got the better of him. One of his students after the First World War, Rudolf Peierls, who would go on to renown as a major participant in the nuclear era through America's Manhattan Project and then British nuclear efforts, recalled Nernst as "a great physicist of rather small stature and an even smaller sense of humility."[14] Perhaps this was due to his incessant need to outdo his rival Arrhenius. But it was also in no small measure tied to his impressive success as a businessman as well as a scientist, something relatively unique among his peers.

In the late 1890s, Nernst had made himself a small fortune with the invention, patenting, and subsequent sale of these patents on a form of incandescent lighting that was superior to the incandescent lightbulb of Thomas Edison fame. Edison's incandescent bulb had a thin, sensitive-to-air filament that caused it to burn up too quickly if not enclosed in a glass bulb and surrounded by a protective gas. Nernst's improvement was to use a thick ceramic material as a filament that could be heated to incandescence without the adverse effects of air exposure. The light emitted by the ceramic element wasn't as glaring and had a longer life. Nernst sold his patents to AEG in Germany and Westinghouse in the United States, making a significant amount of money on the technology, often rumored to be over one million marks. However large the total sale was, an aside that often accompanied the story of Nernst's entrepreneurial success was that "he was the only physicist who had ever signed a contract with an industrial firm in which the advantage was not on the side of the firm."[15] Unfortunately, the Achilles' heel of the "Nernst Lamp," as it was called, was that the ceramic filament took a few minutes to heat up to incandescence. When compared to instantaneous light produced by flipping the switch to a standard incandescent bulb, this seemed like a lifetime, leading to the eventual commercial failure to win over the public to what was a great scientific achievement. Now, Nernst was determined to make sure the conference he had initiated would not have a similar promising beginning, only to fade into the darkness of forgotten history as had his promising lighting invention.

Another prominent attendee was Ernest Rutherford, who was more than pleased to be a part of the proceedings. He had been the chief rival to Marie and Pierre Curie as they had raced to discover the secrets of radioactivity ten years earlier, himself eventually winning the 1908 Nobel Prize in chemistry for his investigations into the disintegration of radioactive elements. Rutherford was a native New Zealander who had spent his adult life in progressively more prestigious positions with English universities. At the time of the First Solvay Conference, he was the leader of a large physics research institute at the University of Manchester. Just months before the congress, Rutherford and his team conducted what would be viewed as a series of perhaps his most famous experiments, "which showed that the atom could be thought of as having a small, highly charged central core, or *nucleus*, surrounded at some distance by a cloud of electrons,"[16] a heretofore unheard-of view of the atomic world. Ultimately, he would move on to take over the esteemed Cavendish Laboratory at Cambridge University after World War I. Certainly, Rutherford's presence in Brussels represented an important opinion to be heard. He was fond of expressing his thoughts to others, usually rather directly, including on the various scientific conferences he periodically attended. In this regard, he wrote to fellow radio-physicist and friend, American Bertram Boltwood, "I am going at the end of next week to Brussels to take part in a small Congress, about fifteen people, on the Theory of Radiations. Some wealthy man in Brussels pays a thousand francs each for our expenses. This is the sort of Congress I have no objections to attending."[17]

At the time of the Solvay Conference, Marie Curie and Ernest Rutherford were really no longer competing on radioactive research, both having made a name for themselves in their discoveries and characterizations of radioactive elements and winning Nobel Prizes for their efforts. For the most part they now, more often than not, worked more in concert to further the knowledge and productive use of radioactivity. In the years since her husband's death, along with her swelling fame as one of, if not the, preeminent scientists of radioactivity, Curie had become defensive, even somewhat prickly, about her stature. But Rutherford, as the other recognized leader in the field, knew the secret of persuasion with her, approaching Curie as an equal but still showing deference to her position.

Just the year before, they had spent some time together during the fall as part of another conference, the International Congress on Radiology and Electricity. The meeting was held in Brussels and partially sponsored by Solvay as part of his wide-ranging scientific interests. More typical for this type of international scientific gathering, close to five hundred scientists attended from around the world. As was often the case, Rutherford's stature and ease of speaking led to his giving a keynote opening address. According to Boltwood in an article he authored at the time on the meeting, Rutherford "pointed out the importance of a uniform, international standard by which the results and experiments of workers in all parts of the world [in measuring radiation strength of radium] might be brought into accord."[18] In short order, the two reigning authorities on the subject, Rutherford and Curie, were tasked with forming a committee at the congress to determine how this standard should be developed. After much discussion, some of it contentious, the committee agreed that Marie Curie would prepare the standard sample for radium and that it would be housed, at least initially, in her laboratory in Paris. Further, upon Marie's stubborn insistence, the unit of measure based on the rate of decay of one gram of radium was to be named after Pierre, thus designated the "curie." Eventually, Rutherford was able to employ his unique powers of persuasion with Curie to convince her that the international standard should be kept at a location other than her own laboratory that provided easier access to the sample for comparison by an international community of radiologists.

Now, as the week progressed, Curie, Rutherford, and others of the group began to appreciate the extent to which Solvay and his assistants had attempted to make this conference, even down to its venue, a place for comfortable and convivial reflection. While in Brussels the previous year for the Radiology and Electricity gathering, Marie had stayed at the Hotel du Grand Miroir, a dated establishment well past its prime that catered to French clientele. By comparison, the grandeur of the Hotel Metropole, with its wonderful mix of artistic styles, luxurious furnishings, and modern conveniences, had a spacious atmosphere accentuated by high ceilings filled with electric radiance, a place suited to expansive, creative thinking. Even mealtime was made special, the German physicist Arnold Sommerfeld making emphatic note to his wife in a letter home about the overly generous dining arrangements and international participants. "We are the guests of M. Solvay, and this includes the meals. No less than five courses at each dinner! It's crazy! . . . Yesterday evening I had a Frenchman on my right and an Englishman on my left, and I was speaking to them both, one after the other!"[19] The international character of the meeting, in a setting that put everyone at ease and provided an agreeable forum for exchange of ideas, opinions, and speculation was exactly what was needed as the group tackled the new frontiers of physics. As part of Rutherford's concise summary of the conference that he wrote a few weeks later for an English scientific journal, he specifically lauded this new twist to scientific gatherings. "The meeting took place under unusually pleasant social conditions, for all the members were staying at the same hotel and dined together. The interchange of views on many problems of modern physics was a feature of the occasion, and led to a much clearer understanding of the points at issue."[20] The relative intimacy that this environment provided was something altogether different and appreciated by those in attendance.

And so the First Solvay Conference on Physics opened on Monday, October 30. Singly or in small groups, the guests proceeded to a rather compact but well laid out meeting room where the sessions would be held. Light wood paneling warmly surrounded the participants, who were rather tightly packed around an oversized, sturdy conference table, each with their accompanying notebooks, reference texts, pencil, and paper in front of them. A stylish electric chandelier hung from the high ceiling, illuminating the moment. A blackboard and papered easel stood against the wall behind the head of the table, available for speakers to employ if they wished to jot down a clarifying formula or two or sketch an experimental setup to explain their points to the listeners.

The distinguished host, Ernest Solvay, sitting at the head of the conference table, welcomed those in attendance. Then he began to summarize his rather unusual thoughts on gravity and matter, a more in-depth copy of which had been sent to everyone a few months before for review. In his presentation, Solvay took pains to note that "his theory . . . is more along the lines of the philosophy of physics rather

A Marie Sklowdowska Curie Family Reunion (Zakopane, Poland, 1899). Standing: Pierre and Marie Curie (extreme left), sister Bronya and her husband Casimir Dłuski (extreme right), brother Jozef (center) flanked by Casimir's two brothers. Sitting (l to r): sister Helena, father Władysław, Casimir's mother. Daughters of Helena and Bronya sitting in front of Władysław. *Courtesy of Association Curie et Joliot-Curie, Musée Curie.*

LEFT: Marie and Pierre Curie in their laboratory (Paris, 1896). They met in 1894 and married the following year. *Courtesy of AIP Emilio Segrè Visual Archives, Physics Today Collection.* BELOW: Mileva Marić and husband Albert Einstein (Prague, 1912). They had met fifteen years earlier as students at the Zurich Polytechnic and married in 1903. *Courtesy of ETH-Bibliothek Zurich.*

LEFT: Ernest Solvay in later life, industrial entrepreneur turned scientific philanthropist. *Courtesy of the Solvay Heritage Collection.* BELOW: Ernest Solvay, over 80, on one of his last Alpine ascents. *Courtesy of the Solvay Heritage Collection.*

ABOVE: Marie Curie (left) and Paul Langevin (right), two attendees from the French contingent at the 1911 First Solvay Conference on Physics in Brussels, Belgium. *Courtesy of the Solvay Heritage Collection.*

RIGHT: Albert Einstein, at 32 the youngest of the invited participants at the First Solvay Conference. *Courtesy of the Solvay Heritage Collection.*

ABOVE: Walther Nernst (left), who proposed the idea of the First Solvay Conference on Physics, and Hendrik Lorentz (right), the incomparable moderator of the conference. *Courtesy of the Solvay Heritage Collection.*

LEFT: Marcel Brillouin, part of the French contingent at the First Solvay Conference and a confidential chronicler of the affair between Marie Curie and Paul Langevin. *Courtesy of the Solvay Heritage Collection.*

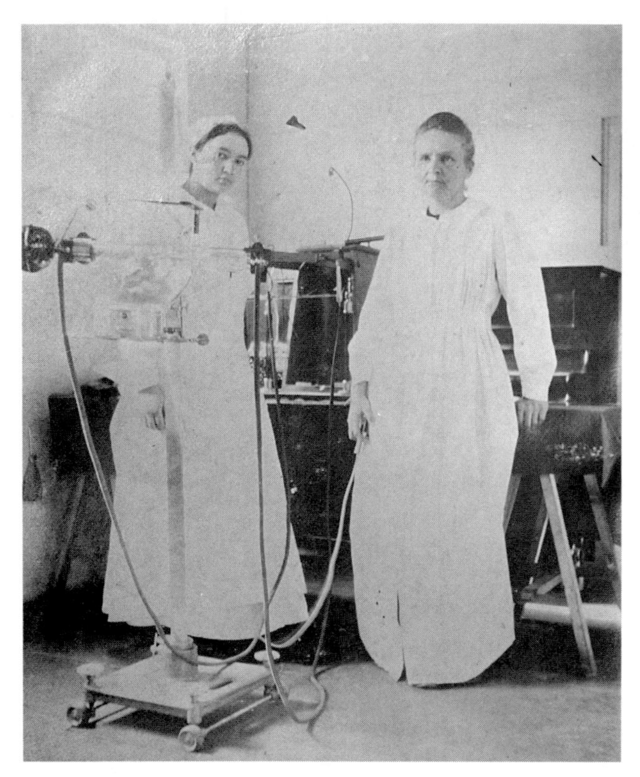

LEFT: Irène Joliot-Curie and her mother Marie surrounded by X-ray machines installed in a hospital in Hoogstade, Belgium during WWI (1915). *Courtesy of Association Curie et Joliot-Curie, Musée Curie.*

American Red Cross X-ray ambulances in World War I used technology made possible by Marie Curie, which were especially useful in detecting shell fragments in the wounded (May 18, 1918). *Courtesy of the Library of Congress, American National Red Cross collection, Prints and Photographs Division.*

RIGHT: Marie Curie and her daughters arrive in New York Harbor via ship for their first ever visit to America (May 1921). Standing (l to r): Trip sponsor Marie Mattingly "Missy" Meloney, editor of the *Delineator* magazine, Irène Joliot-Curie, Marie Curie, Eve Curie. *Courtesy of the Library of Congress, George Bain Collection, Prints and Photographs Division.*

LEFT: Albert Einstein and his second wife Elsa arrive in New York Harbor via ship on their maiden voyage to America (April 1921). *Courtesy of the Library of Congress, George Bain Collection, Prints and Photographs Division.*

Often referred to as "The Most Intelligent Picture Ever Taken," this photograph of the 29 attendees of the Fifth Solvay Conference on Physics contains 17 actual, or soon-to-be, awardees of the Nobel Prize in physics or chemistry (1927). *Courtesy of the Solvay Heritage Collection.*

than standard physics,"[21] a difference that, based on his status as a self-taught physicist at best, provided some wiggle room for his hypotheses versus accepted theory. What Solvay proceeded to review were some of his long-held beliefs that were delivered with utmost sincerity but lacked the scientific rigor expected by the audience. The conference attendees listened politely, but only silence greeted the speaker once he completed his remarks. In this manner, a message was delivered to Solvay by his esteemed guests: thank you very much for inviting us here and sharing your ideas, but it might be best to leave the scientific technicalities to the professionals.

After a pregnant pause, Hendrik Lorentz, sitting nearby and wanting to move on, cleared his throat and quickly thanked Solvay for sponsoring the meeting. He respectfully mentioned the extensive thought Solvay had put into his opening comments but felt it was now time to shift the conversation to the task at hand. He noted the issues before them: "We currently have the feeling that we have reached an impasse. The old theories have proved themselves less and less capable of illuminating the dark shadows that seem to surround us . . . the beautiful hypothesis of elements of energy proposed first by Planck and then applied to numerous problems by Einstein, Nernst, and others, has come as a precious ray of light."[22]

Solvay's invitation letter to the participants mentioned that a number of the attendees would be giving formal presentations on a few key subjects. Copies of these presentations had been forwarded to the complete group, similar to Solvay's own opening remarks, so that the conference members would have the opportunity to familiarize themselves with the material to be discussed. This was another key distinction between the Solvay Conference and the other scientific forums of the day, which had so many speakers covering a multitude of topics such that advanced preparation of this sort would have been out of the question.

In all, twelve presentations beyond the opening remarks of Solvay would be given at the conference during the week. Although they varied a bit individually, all but three could be divided into two main categories: those that addressed radiation or energy, and those that addressed a concept called "specific heat," which referred to an amount of heat as energy that was required to raise the temperature of a mass by one degree Celsius. This was a physics definition that gave a real world indication of how much energy would be needed to raise or cool the temperature of an object by a certain amount, important in understanding the effects of different forms of energy and their reflection, absorption, or emission from solids in experimental situations. In reality, these two topics were both tied to the central issue around which the conference had been created—the mysterious, discrete packets of energy called quanta.

Interestingly, of those who attended, it was the French delegation that was the most poorly acquainted with this common theme being discussed, the new quantum hypothesis concerning light and energy, the major exception being Paul Langevin, who appeared to have kept current especially with Einstein's publications. During the years following Planck's and especially Einstein's work and more recent published articles on the subject, unless one's own research touched specifically on an area closely related to these issues, it was entirely possible to be virtually unaware of what was going on related to early quantum theory. It was understandable that this might be the case for an experimental physicist like Marie Curie, whose efforts weren't as focused on theoretical modeling of various interactions in the natural world. However, even those unfamiliar with the energy quanta had been invited to attend this special conference on radiation and the quanta for their incisiveness and the individual insights they could bring to this scientific conundrum. As well, in keeping with Solvay's insistence on a broad international composition of conference participants, he and Nernst had seen a realistic need to balance out the heavy German-speaking invitee list, who were by comparison steeped in exploration of the quanta, with an equal infusion of French thinkers.

Lorentz, who could not avoid letting Nernst express a few introductory words himself related to the puzzling issue at hand, then offered the first of the prepared papers on the subject, concerning some very well-accepted theories on radiation. The report pointedly gave a number of "reasons to believe Planck's radiation formula [involving quanta], which had withstood every experimental test, was incompatible with ordinary physics."[23] Many in the audience took part in the discussion following his presentation and generally agreed with this sentiment.

Typical of the group's exchanges was when Englishman James Jeans spoke next on specific heats, supported by a letter from another English physicist, Lord Rayleigh, who could not attend the meeting. Jeans's intent was to lend support to classic physics interpretations of energy, but vigorous debate was caused by his explanation of some energy phenomena that required an underlying belief in a version of equilibrium theory that was not justified by experimental evidence.[24] Difficulties with Jeans's approach were expressed by a significant faction of the scientists and were capped off by Poincaré's insistence that Jeans was introducing too many parameters to explain his results appropriately. Poincaré stated, "This, however, is not the role of physical theories. They should not introduce as many arbitrary constants as there are phenomena to be explained; they should establish a connection between the various experimental facts and above all permit predictions."[25] Although Jeans defended his views on classical physics, he generally offered little resistance to the idea that quantum theory might actually present a more realistic approach to some issues, eventually stating, "However, I hardly believe in the prospect that the classical theory, together with some other new hypothesis regarding the mechanism of radiation . . . will ever lead to formulae that will reflect the facts as well as Planck's equations [on energy quanta]."[26]

Mark Knudsen of Denmark gave remarks on the kinetic theory of ideal gases, the concept that gases are composed of a myriad of small molecules in constant motion and colliding with each other, the motion increasing as the temperature increases. Germans Warburg and Rubens gave related papers supporting Planck's experimental efforts on energy, but qualified their conclusions by saying that more evidence needed to be developed to substantiate Planck's theory. Sommerfeld presented on the applicability of quantum theory to atoms and molecules, while Nernst himself focused on how his own experimental efforts at low temperatures were more quantitatively correct when employing quantum theory than classical calculations, in agreement with Einstein's work on specific heats and various solids. Perrin, who discussed molecular theory; Kamerlingh Onnes, whose topic was electrical resistivity; and Langevin, lecturing on magnetism, covered scientific concepts that were more tangential to energy and the quanta.[27]

Planck, of course, covered the reason that the conference had been initiated, his own experiments on heat radiation and resulting conclusions that had driven him to the existence of the discrete packets of energy termed quanta. Only now, his conservative nature had gotten the better of him. He elaborated on a new twist he had thrown into his observations, hedging somewhat on his original concept by stating that "emission of radiating energy was considered to occur in quanta, whereas adsorption was relegated to the prior 'classical' continuous process."[28] It seemed that Planck was still uncomfortable with the radical thinking that his own theoretical creation of the energy quanta had unleashed upon physics.

Likewise, in the last paper given at the conference, Einstein dealt with quantum theory, especially in low-temperature environments.

As he reviewed his work, he offered that "the theory of quanta, in its present form, can be useful, but that it doesn't constitute an actual theory in the usual sense of the word . . . On the other hand, it has now been established that classical [physics] . . . cannot be considered a generally useful scheme for the theoretical representation of all physical phenomena."[29] Einstein, as gently as possible, was indicating what many felt—that something was deeply amiss in classical physics theory.

The radioactivity experts, Curie and Rutherford, did not present papers at this conference. Instead, they absorbed the information presented by others, reflected on the theories as well as the speculation, and commented appropriately in the discussions following the formal presentations. According to biographer Quinn, Curie was "a lively participant in the discussions, questioning, conjecturing, and adding information from her own work. At one point, she engaged in a lengthy discussion with Rutherford on the nature of [radioactive] decay."[30]

The intense focus that Curie brought to the proceedings was clearly on display in the classic group photograph of the First Solvay Conference attendees that was taken for posterity. In it, she and Poincaré can be seen seated at the conference table, on the right-hand side in the picture. They stare intently at a text of some sort, Marie almost looking like a studious pupil, head-in-hand as she examines the text while her professor, Henri, provides guidance. Both appear lost in concentration while the rest of the assembled guests politely look at the camera. A few even offered a smile, like Ernest Rutherford, who towered directly over her shoulder in the picture, his face adorned with a self-satisfied grin. But it almost seemed that Curie, as the only woman scientist present, didn't want to be viewed as taking time out to relax, even for a picture, while in the company of her male counterparts.

When Curie and Rutherford had last met the previous autumn in Brussels, Marie had not been well. Rutherford remembered he had actually escorted Curie, complaining of exhaustion, home prematurely from an opera that many conference participants had attended one evening during the proceedings.[31] Now, he saw she was in better health and spirits. She mixed with the group of scientists, meeting people who to her had only been names in scientific journal articles the week before, exchanging pleasantries with those she had met at previous conferences, including the Dutchmen Lorentz and Kamerlingh Onnes, and of course Rutherford. As noted in Robert Reid's biography of Curie, he states that Rutherford "had heard, along with other delegates to the conference, of the rumors spreading in Paris about some business between her and Langevin and breezily dismissed them as 'moonshine.'"[32] Even though Rutherford said he paid no heed to the discreet whispers, there was a silent undercurrent that ran through the flow of the meeting. The formal discussions were the large, visible tip of an iceberg that constituted the conference. A significant portion, submerged under the waterline, was composed of the personal interactions at close quarters of so many brilliant personalities, including the unspoken circumstances of Marie Curie.

Many of the scientists had never met before, except perhaps at the few international congresses that were just starting to take place. There, hundreds were present, affording little chance for anything but formal greetings or superficial encounters. Here in Brussels, someone like the twenty-five-year-old Frederick Lindemann, a laboratory assistant of Nernst's who had come along to act as a secretary to the proceedings and would, a generation later, become a scientific advisor to Prime Minister Winston Churchill, got the chance to meet an almost-equally youthful thirty-two-year-old Albert Einstein. Lindemann's father was German but his mother was from a wealthy family in England. Though his conversations with Einstein left a profound impression with Lindemann, Frederick's ingrained class bias was clearly evident as he wrote a letter to his father from the conference, "[Einstein] asked me to come stay with him if I came to Prague and I nearly asked him to come see us at Sidholme. However, he does not care much for appearances and goes to dinner in a frock coat."[33] It seems that nearly everyone in the group wanted to get to know Einstein in greater depth as, along with Max Planck, he was one of the two leading proponents of quantum thought. Planck and Einstein themselves, in addition to being the de facto leaders of the quanta queries, shared a devotion to musical performance, the former being an accomplished pianist, the latter having a proficiency with the violin. This perhaps led to some lighter discussion between the two on the potential for a future duet.

Nernst already was favorably acquainted with Einstein, having actually visited him the year before to discuss low-temperature phenomena he was investigating and its connection with quantum theory. As to Curie, he had supposedly first met her years ago, not at a conference but at a reception, shortly after she had become famous for her discovery of radioactivity and radium. The story goes that "Nernst and [Lord Kelvin] met Mme Curie who in her young years was extremely good-looking. They asked her about the newly discovered radium and she told them that she had brought a sample along, but that the lights were too bright to see its glow. So the three of them squeezed into the space between double doors, but before their eyes had grown accustomed to the dark, there was a knock at one of the doors. It was Lady Kelvin, who thus showed that she was, as Nernst put it, 'a most attentive spouse.'"[34] This early contact between the two had probably left no impression on Curie at all, but it seemed a fond anecdote available for Nernst to periodically trot out on occasion. One could envision him discussing the incident at one of the after-dinner sessions at the conference, over brandy and cigars with a select few newfound scientific comrades.

Although she had aged considerably in only a decade due to her constant exposure to radioactivity coupled with the always-present loss of her husband, Marie had a quiet charm and still could exhibit a vibrant personality when she had the energy and desire to do so.

At the conference, she and a few of her French cohorts were able to get to know Einstein. Curie was probably at the height of her fame at this point, having won her Nobel Prize and various other international honors a number of years before. This had been followed by her devotion to furthering research on radium, its characteristics, and its promising potential for medical use in treatment of certain cancers. During these years, Curie had compiled and recently published what was the definitive text on radioactivity, a meticulously detailed effort of almost one thousand pages that was universally admired by the scientific community and specifically by Rutherford and many others in the field. She had even been able to painstakingly isolate metallic radium, versus the radium salts that were the standards employed in its use, a time-consuming, labor-intensive feat that was impressive in its own right.

In contrast, Einstein was the youngest and perhaps the least-known of the specially invited assembly of genius at the congress. His earlier theoretical pronouncements had mainly been published within the last six years, proclaiming his innovative brilliance but to a decidedly smaller slice of the world of physics. Of those in attendance at the Hotel Metropole, the German-speaking contingent at the conference were those that knew him best. Curie had most likely encountered Einstein in 1909 at the 350th anniversary celebration of the founding of the University of Geneva, when they both had been selected to receive honorary degrees along with Hendrik Lorentz and Ernest Solvay. But there were over one hundred others being so feted, making anything beyond initial introductions unlikely. It was at the First Solvay Conference on Physics that they were to form the beginnings of a lasting friendship.

Each observed the other, both in the formal sessions as well as during the planned allotment of time at the end of the day for dinner and more casual conversation. They were immediately impressed by, and attracted to, the thought processes of each other. Einstein referred to

Curie's "sparkling intelligence," while during the week Curie perceived enough of Einstein's vast intellectual capacity to feel comfortable describing his mental acuity in a glowing reference she provided to the Swiss Federal Institute of Technology in Zurich shortly after the conference when he sought a professorship there. Discussions of family mixed with science, making for more personal conversation. Marie had two girls, the studious, intellectual, fourteen-year-old Irène and Eve, half the age of her older sister. Eve was slightly younger than the oldest of Albert's two sons, seven-year-old Hans Albert and a baby brother, Eduard. The older children enjoyed outdoor hiking, as did the two scientists, who discussed the possibility of arranging a future excursion of some sort with the families. A few years later, this trip actually took place, a hiking holiday in the Swiss Alps remembered by Eve Curie as part of the "charming comradeship of genius" that had existed for the past two years between them, built on mutual admiration and a friendship that "was frank and loyal."[35]

The week proceeded in this fashion: daily presentations, discussion, and at times intense debate infused with a collegial atmosphere being promoted in large part by Lorentz. Fifteen years later, Marcel Brillouin, in a journal article on Lorentz, remembered the sessions being guided in the Dutchman's inimitable, multilingual fashion, "with ceaseless actions to prevent discussion from getting off track . . . how easy it would have been for the little room in the Hotel Metropole to become a true Babel! But Lorentz followed everything, stopping speakers if their speeches, a little difficult to comprehend, left some of us a little puzzled, to summarize the essentials, distilled by his clear intelligence, in the two other languages of participants."[36] A quick lunch at midday, then following the afternoon's work another sumptuous dinner with the whole group in attendance, international personalities continuing the discussion of scientific phenomena, interspersed with personal reminiscences and good-natured stories.

In a blink, midweek had arrived, and with it an unexpected telegram for Curie that provided her with an extraordinary feeling of satisfaction. The Nobel Prize Committee had decided to award her another honor. As opposed to her first award, which was in physics for discovering the phenomenon of radioactivity, this one was in chemistry, for the discovery of the radioactive elements radium and polonium. And, pointedly, this one was for her efforts alone, shared with no one else, not even Pierre. The message was an advance notice to her being provided by the Committee, who would send a more formal, congratulatory message on the award the following week, to be publicized appropriately. Although Marie was thrilled to receive the notice, secrecy of award winners was to be strictly enforced and it appears she kept the good news to herself, probably only sharing it in confidence with a select few of the French delegation, at the very least with her special friend Paul Langevin and possibly her former Parisian neighbor Jean Perrin.

Even though the awarding of annual Nobel Prizes was only a decade old in 1911, it was steadily becoming a pinnacle of scientific honor. It was coveted, but not necessarily sought, by most everyone in the disciplines of science that were eligible for this distinction. An award bestowed was considered major peer recognition, to some even a lifetime achievement. In millionaire industrialist Alfred Nobel's will he had specifically indicated that the Royal Swedish Academy of Sciences would be in charge of the prizes for physics and chemistry. This body accomplished the task of selecting award recipients through the Nobel Committees for Physics and Chemistry, a handful of Swedish university professors in physics and chemistry who culled through letters of nomination sent in by qualified individuals of science. They developed a list of finalists, assigning specific professors on the committee to perform detailed assessments on these candidates and report back so that comparisons by the full team could be made between the various nominees to arrive at an annual winner.[37] In doing so, the committee needed to be up-to-date on current technical advancements in the field as well as attempt to be as objective as possible in assessing various finalists. Swedish scientific giant Svante Arrhenius was an influential member of the Physics Committee, a past recipient of the Nobel Prize for chemistry, and generally a powerful force within the Swedish academicians deciding the fate of the award nominees. As such, his continuing professional feud with Walther Nernst had forced the latter to become more creative in his approach to obtaining an award of his own, in large part leading to his concept of the Solvay Conference.

At this point, Marie Curie had been the only woman privileged to win a Nobel Prize, in 1903 sharing the honor with her husband Pierre, as well as Henri Becquerel, for their work on radioactivity. No one had ever won more than one, that is until Marie Curie received her special delivery telegram during the conference. In 1903 and for years afterward, many had questioned Marie's actual involvement in the efforts of the two Curies related to radioactivity, unwilling to believe that a woman could work as an equal with a man in scientific thought related to this type of breakthrough. Surely, she was a helpful assistant, they imagined, but it didn't seem appropriate in their reasoning to present her with a Nobel Prize for what must have been a man's ingenuity in discovering the natural wonder of radioactivity.

For Marie to win a second award was almost incomprehensible, in spite of all she had done in the intervening years to demonstrate her intellect and further the scientific knowledge in this arena. Of course, there was nothing new in this claim of women's inferiority in the world of science. Curie's English contemporary and friend Hertha Ayrton had come up against the same prejudgment when both she and her husband Professor William Ayrton performed electrical experiments in concert. Their remedy was to consciously devise ways to circumvent this situation, "[William insisting] on her own line of research whenever possible, and making her own experiments. He knew well enough how readily any success that she achieved would be attributed to the assistance of her husband . . ."[38]

To some, it appeared that Curie was unfairly receiving two awards for the same thing. The 1903 award for the discovery of the occurrence of radioactivity could be thought of as an overarching achievement that also covered the discovery of the distinct radioactive elements of polonium and radium. A 1996 lecture given to the Royal Academy of Sciences by Swedish physicist Nanny Fröman disputes this contention, stating that "the citation for the Prize [in physics] in 1903 was worded deliberately with a view to a future Prize in chemistry. Chemists considered that the discovery and isolation of radium was the greatest event in chemistry since the discovery of oxygen. That for the first time in history it could be shown that an element could be transmuted into another element, revolutionizing chemistry."[39] The Nobel Committee obviously felt the all-encompassing award argument was not applicable. Nevertheless, this feeling, combined with the initial protestations of Curie's unworthiness to receive a Nobel Prize in the first place, became just another gathering wave in the tsunami building against Marie Curie that was to explode with tremendous fury in the French press over the coming months.

For, just as Marie was reaching a professional height never before attained by a physicist or chemist for being awarded her second Nobel Prize, she was sent reeling with a trailing telegram she received from another source, hard on the heels of the committee's message, informing her that Paul Langevin's wife was about to publicly announce a secret love affair between Curie and Langevin to attempt to smear her as a miscreant to honored French values of family and country. As the conference came to its conclusion, the devastating article on Langevin's suspected infidelities with a conniving Madame Curie appeared on the front page of *Le Journal*. It forced a hasty retreat from the meeting in

Brussels as Curie and Langevin took the next available train back to Paris on Saturday.

As to the Solvay Conference itself, feelings were mixed. The meeting definitely established an international recognition among the group of the perplexing questions facing physics. Some said that although the council "aired everything, it settled nothing," in many ways leaving the quanta more puzzling than ever.[40] Einstein was convinced that many in attendance just didn't understand the significance of the principles being proposed, not the least of whom was Poincaré, who Einstein thought was too grounded in classical theory to accept a different approach.[41] To the contrary, the idea of the quanta had made such an impression on Poincaré that following the conference he quickly immersed himself in its study. He soon pronounced that "the hypothesis of the quanta is the only one which leads to Planck's [conclusions]," lending his gravitas in favor of the argument.[42] Marcel Brillouin perhaps put it best when he commented on the new, discrete packets of energy versus its formerly accepted continuous nature. He stated that "It seems certain that from now on, we will have to introduce into our physical and chemical conceptions the notion of discontinuity, some principle of variation in jumps, which we simply had no idea of a few years ago."[43] But all could easily conclude that a new era of thought in physics was beginning.

The form of the conference was unlike anything the scientific world had experienced, and it served its purpose admirably. The small contingent of learned individuals, focused on a few singular issues, had been able to clearly state and consider the problems, if not arriving at substantive conclusions. What became vitally important now was the need for the establishment of a means to construct and execute experimentation that could work toward solving the riddles discussed at the conference. Within months, Hendrik Lorentz was working with Ernest Solvay on the founding of the Solvay International Institute of Physics, initially bankrolled with one million francs from the ever-generous benefactor Solvay. The basis of scientific inquiry for the institute would be worthy projects as judged by a committee initially composed of members of the conference attendees. Markedly differing from Nobel Prizes, for which recipients were give cash awards for work already accomplished and judged uniquely impactful in science, the Solvay Institute would provide monies for experiments yet to be performed, focused on solving problems deemed by the committee to be of utmost scientific importance.

At the same time, the novel gathering at the Metropole showcased a shining new talent that lay in the ruminations of a recently established professor of physics at the University of Prague. Albert Einstein had demonstrated the mind of a superior critical thinker, a complex, multifaceted mechanism that was both respected and envied by those witnessing its functioning. Thanks to this debut, he would shortly leave his post in Prague to return to Zurich as a professor there, only to pull up stakes again a year later for a truly impressive professorship carved out specifically for him by the Prussian Academy of Science, lobbied for courtesy of effusive recommendations from Planck, Nernst, and others from the German faction at the conference praising Einstein's impeccable credentials.

Just as importantly, the First Solvay Conference on Physics had become the setting for the establishment of what was essentially the forging of a lasting bond between Marie Curie and Albert Einstein. Amid the rushing chaos of the conflict between the theories of classical physics and the enigma of the quanta, the two had taken the opportunity to become acquainted in a manner that allowed them to glimpse the other's human sensitivities as well as their brilliance. As the firestorm of the sensationalist French press was about to engulf Marie, tearing from her grasp a last chance at love, the simple, loyal support she was to receive from Einstein was surely a comfort in a world she would rightly say had gone so terribly wrong.

CHAPTER EIGHT

Marie Curie's Impossible Dream

The Louisiana Purchase Exposition of 1904, informally termed the St. Louis World's Fair, celebrated the one hundredth anniversary of America's purchase of the Louisiana Territory from the French. The event took place from April through December and, though nothing on the scale of the 1900 World's Fair in Paris, it attracted close to twenty million visitors to the heart of the United States. Two such guests were Henri Poincaré and Paul Langevin, the incomparable French mathematician and his young associate, who came to give presentations at the Congress of Arts and Sciences that was held in conjunction with the fair.

On September 22, in the physics section of the meeting's agenda, Langevin was the first speaker. He would be followed by Ernest Rutherford, lecturing on radioactivity, certainly indicating Langevin was in rarefied scientific air. He gave a paper on the physics of electrons and how this theory applied to other scientific disciplines. J. J. Thomson had discovered the revolutionary particle known as the electron in the late 1890s and subsequently Langevin investigated the phenomenon himself as part of Thomson's ongoing efforts at Cambridge University's Cavendish Laboratory. In his lecture's introduction, Langevin indicated that questions about the theory of electrons existed that needed answers, much like, as he put it in an early paragraph of his paper, a "New America, full of wealth yet unknown, where one can breathe freely, which invites all our activities, and which can teach many things to the Old World."[1] The metaphor was prescient concerning understandings yet to be developed about the atomic world, not the least of which would be depicting the coming collision of newborn quantum thought with classical physics theory.

Paul Langevin was a scientific talent steadily on the rise. Five years younger than Marie Curie, she had known him for many years before they would become lovers. Pierre Curie had been Paul's respected, if not worshipped, professor and mentor at the EPCI in the mid-1890s. Langevin next studied at Cambridge under Thomson, then spent the first few years of the new century researching magnetism. He combined his studies on para-magnetism and temperature effects that Pierre Curie had uncovered with the electron theory he had learned under Thomson to reveal that the spinning motion of electrons caused the magnetic behavior that Curie had observed, an important physics breakthrough. He succeeded Pierre as an EPCI physics professor when Curie moved into his own professorship at the Sorbonne in 1905. During that period, Langevin explored a number of the topics that intrigued Einstein as well, resulting in Langevin's unpublished view of the special relationship of mass and energy that led him to "the idea to associate a mass . . . to the energy of [quanta] or electrons," for which Einstein was to become famous.[2]

Langevin had an inquisitive, imaginative mind paired with a caring character. He was tall and thin, with a combination of hair on his head that blended into a rather striking look. A tiny goatee under his lower lip was overwhelmed by a healthy handlebar moustache, the image completed by a Prussian general's haircut standing at rigid attention. His dark eyes were almost mesmerizing in power, providing a distinct magnetic effect all his own, certainly having something of that result with Marie Curie.

Curie was still devastated by the sudden loss of her beloved Pierre. Together they had conquered the world of physics with their shared efforts on radioactivity and had built a family. She still periodically dwelled on the life partner she had lost and couldn't help but be revisited by the overwhelming depression she had first suffered when her mother and sister had passed away. But she was determined to go on, if only for the sake of Pierre's memory as well as for her two precious daughters. For their benefit, she joined with a few of the other university faculty who had children of school age, including the families of the Perrins and Langevins, to create a unique home-schooling experience for them. During the two years of its existence, this learning experiment boasted the likes of Jean Perrin teaching physics to the children, his wife Henriette focusing on history and French, Paul Langevin presiding over mathematics, and Marie administering the chemistry lessons.[3] As one can imagine, this brought the adults into close proximity on almost a daily basis, with time to get to know each other better. Professionally, Marie immersed herself in her own professorial duties at the Sorbonne and continued explorations on radioactivity. But for herself, it was a lonely existence, the days listlessly turning into months, then years, Marie sometimes wondering if she would truly experience companionship ever again.

At the same time, the union in 1898 of Paul Langevin and his wife Jeanne had been disintegrating in a roller-coaster fashion, periods that built to heated arguments interspersed with stretches of relative calm, even contentedness. The couple appeared mismatched almost from the start, Paul immersed in his science while his wife and her working-class family complained that he did not take advantage of moneymaking business opportunities over scientific pursuits. Not atypical marital disagreements about in-laws and finances were common, but as time passed these began to include physically abusive incidents. At one point, Paul was confronted by his wife and her mother in an argument that devolved into the women flinging iron chairs at Paul, producing bruises on his face that Langevin tried to explain away to fellow professors as being caused by a fall from his bicycle.[4] In another instance, Langevin claimed his wife had broken a bottle over his head and showed the scar to prove it.[5] For his part, Paul had once slapped his wife at the dinner table, Jeanne stating it was because Paul said she had improperly cooked a dessert, he insisting it was as a reaction to her shouting insults at him in front of the children.[6] Through all of this, Jeanne suspected that Paul had occasionally been unfaithful to her, but she had tolerated these indiscretions in a "hidden-in-plain-sight" tradition of the French, whose culture simply looked the other way in allowing married men to keep mistresses as long as they were not paraded in public for all to see. The Langevins stayed together in this codependent manner, both at times threatening to leave each other only to reconcile and resume, literally battered but not broken. Four children managed to be produced in ten years of the marriage, evidencing some periods of truce if not love.

Marie and Paul, each yearning for something more than their present condition, found their friendship transforming into true affection. Marie could envision a future with Paul. He was someone who was as dedicated to science as herself, and perhaps they could even re-create the wonderful pairing that she had made with her former husband. He, on the other hand, appeared to have a shorter-term view, seeking emotional as well as physical comfort from Marie to counterbalance the increasingly turbulent confrontations of his homelife, seemingly content with another mistress rather than a new missus. Where Marie was unbreakably resolute, unwavering in her attempt to find a new life with Paul, he was inherently weak, swayed by feelings of the moment. This would be the tragic flaw in their connection, leading to its ultimate downfall. As the Langevin marriage continued its collapse, Marie and Paul became closer. Curie offered a sympathetic shoulder at first, then became more entangled as she began to picture a future for them together. Langevin proceeded to rent a two-room apartment in Paris, a convenient walk from the university and Curie's laboratory. Meetings there became more frequent, both at midday and overnight.

Marie had had three significant romantic relationships with men in her life, but only her marriage to Pierre was to be the fulfilment of her deservedly high expectations of love. The first was with Casimir on his family's estate back in Poland when she was a young governess. It was a first love with no real plans beyond their infatuation with each other. Now, left with a solitary existence and the memories of her dead husband, she was captivated by Paul Langevin. But except for Pierre, Marie had chosen poorly. More so than Casimir before him, Paul had a promising future but with similarly little determination when confronted with challenges to his relationship with Marie, circumstances that needed stalwart action to surmount any obstacles. Curie clearly recognized Langevin's lack of resolve, complaining to another woman whom she trusted that while both women were strong, Langevin was not, requiring understanding and warmth to persevere.[7] So, when his devotion to her was finally put to the test, it came as little surprise that Langevin failed in rather spectacular fashion, a sorry ending to Curie's impossible dream.

Marie and Paul were two of the scientific professors of the Sorbonne and the College of France, institutions across the street from each other and both at the highest level of education in the country, supporting a small, intimate community of excellence. By the turn of the century, the Langevins were acquaintances with the Curies, the strong connection between the four adults resting primarily on Pierre's teaching and mentorship of Paul. The two men appeared alike in many ways, from being gifted scientists to their progressive politics. Marie's subsequent blossoming interest in Paul almost seemed a logical progression from their initial friendship and more frequent interaction after Pierre's death.

As well, the Langevins were friendly with a number of other couples whose husbands taught at the Sorbonne or the College of France. Paul and Jean Perrin were both young stars in the world of physics, their wives seeming to get along well, Jeanne making Perrin's wife Henriette a confidante. Marcel Brillouin, a professor at the College of France, and his wife Charlotte were also quite friendly with the Langevins. Brillouin was a prominent mathematical physicist, a talented teacher who authored many papers on various scientific topics but rarely garnered awards for his work. Still, he was important enough to be included on the list of French attendees Ernest Solvay and Walther Nernst were to construct for the invitations to the First Solvay Conference on Physics. Marcel was almost twenty years older than Langevin and thought of himself as more of a teacher than a peer to Paul, although both had been professors together for a number of years at the same institution.

It was based on this friendship, and the information that Jeanne began to pass along to Charlotte in early 1911 about her marital issues, that Marcel began to keep an extensive journal on the lamentations that he and his wife were hearing from Jeanne Langevin and her brother-in-law Henri Bourgeois concerning Paul and Marie. It, as well as some accompanying correspondence among Jean Perrin, Hendrik Lorentz, and Marcel, provides what appears to be an heretofore unexplored perspective on the Curie/Langevin relationship from the vantage point of someone not of Marie Curie's inner circle of friends but who was acquainted with the Curies and had significant standing in the Parisian academic community. The papers represent a departure from the more sympathetic support of Marie previously published from her friends' writings about this volatile time in her life. They conveyed a conservative, probably more typical Frenchman's cultural position on the evolving relationship. In this vein, they were fairly uniform in their stern disapproval, offering little understanding or sympathy for Marie Curie's situation. Unfortunately, this would be an attitude that would be prevalent in much of Paris, and across France, as the affair became exposed to all.

One of the earliest journal entries described a visit by Jeanne Langevin to see Charlotte Brillouin on March 11, 1911, confessing that her marriage was in deep trouble. Marcel wrote: "For two years, Langevin has been spending all his time with [Marie Curie] . . . It seems that Mrs. Curie sports white dresses, etc.—Langevin, who does not want to take his wife or his children bike riding any more, goes off with Mrs. Curie alone, etc. etc. . . . He does not want to go to Congresses with his wife, because Mrs. Curie is going . . . All of this is the exact opposite of the Langevin of the early years of his marriage when he was so devoted to his wife and children."[8]

The reference to Curie sporting white dresses signified the noticeable change that had come over Marie, who habitually had worn black dresses, especially since the death of Pierre. Various biographers noted that other friends of Marie's mentioned the detail of the white dresses in their remembrances as Curie's expression of newfound happiness. Here Brillouin contrasted the white dresses with a particular grey one she wore later that evening. He described a gathering attended by the Brillouins and hosted by a well-known French chemist of the Sorbonne, Albin Haller, where Curie made what appeared to Marcel a most alarming entrance: "Mrs. Curie arrived at the dinner party dressed in a ridiculous fashion unfitting for the widow of Curie, grey dress, bodice open to the bust line, arms bare to the neck, transparent sleeves, branches of flowers to the bust, all this bringing out her age and her thinness, seductive attitudes—impression that I regret to find in accordance with Mrs. Langevin's conversation [with Charlotte]."[9]

Obviously, Brillouin was taken aback by Marie's lack of continuing to mourn her dead husband in what he felt was an appropriate manner, dressing in a fashion he found most unfitting for the widow of the sainted Pierre Curie, who at this point had been dead for almost five years.

A few weeks later, on Wednesday, March 29, Jeanne visited Charlotte again, this time accompanied by her brother-in-law, Henri Bourgeois, an editor of a Parisian newspaper, *Le Petit Journal*. They described to Charlotte the continuing developments of the affair, as Marcel detailed in his journal: "Langevin had been ill with an abscess for several days, Mrs. Langevin found a whole lot of letters from Mrs. Curie, perfectly explicit—it is Mrs. Curie who forbade him to go to his wife's bedroom, urged him to stop giving her money for the household, to leave her, etc., until she files for divorce . . . Hesitating to abandon his children, [Langevin] receives the advice that he will forget them and anyway, Mrs. Curie needs a father for her children, etc., an abominable, practical series of pieces of selfish advice to which [Langevin] weakly obeys."[10]

Brillouin went on to note that they had rented rooms in Paris, "in short, it is impossible to doubt that Mrs. Curie is passionately in love with Langevin, has given herself to and imposed herself on him and dominates him . . . It is quite obvious that Langevin could never have desired Mrs. Curie . . . and moreover had such admiration and veneration for the memory of [Pierre] that one supposes that he must have resisted for a long time."[11]

Brillouin referred here to Jeanne Langevin's discovery of the infamous personal love letters between the two, which contained intimate, loving exchanges between Curie and Langevin. Also included in them was Curie's rather practical but self-serving and sometimes insensitive advice and how Langevin submitted to her requests. He noted that she was no doubt in love and had overwhelmed Langevin, Marcel giving no thought to the nature of Marie's predicament in fighting for the very existence of a love that she desperately desired but about which Paul seemed interested but often ambivalent. In Brillouin's way of thinking, Paul would never have been able to besmirch the memory of the great Pierre Curie, making it obvious that Marie had thrown herself on him. This appears to be at odds with numerous indications that have been detailed by others about Langevin's statements concerning the relationship, including that he sought from Marie "a little tenderness which I missed at home,"[12] and that, writing in notes to Marie about a meeting at their rented apartment in Paris, "my darling, I will not stop thinking of you. I embrace you tenderly."[13]

As Brillouin had jotted down in his journal, both he and his wife were shocked and dismayed by what Jeanne conveyed related to the contents of the love letters. He went on to outline a visit by Henri Bourgeois to him the next day at the College of France, explaining Bourgeois's desire to separate Langevin and Curie without causing a public scandal. Bourgeois mentioned the letters again, Brillouin listening and advising to do nothing further with the letters but offering no further assistance, the journal entry for that day noting: "I had thought of intervening with Perrin when I thought we were at the beginning of a love affair, and not at crisis level . . ."

"Talking to Langevin would seem impossible to me. I would risk saying what I think of both of them with too much vehemence; and if [Langevin] came to understand and committed suicide on leaving my home, I will mean that I have, unfortunately, gone too far."[14]

Brillouin felt he could go no further than to listen to Jeanne and Henri, at this late date in the progress of the affair feeling that combining forces with mutual friend Jean Perrin to address the issue would also not be useful. Marcel felt that if he was pushed to directly intervene by confronting Langevin, he might convey to Paul such a blistering judgment of distain about the affair that it might cause a response that Marcel could not chance, leading to Langevin possibly taking his own life over the embarrassment and dishonor of it all.

The entry for this meeting closed with Bourgeois dangling the prospect of scandal for the respected College of France as a potential result of the exposure of the Curie/Langevin relationship. Brillouin counseled that this result would only lead to so many people finding out about the affair that the relationship between Langevin and his wife Jeanne would be shattered forever, something Bourgeois said he wanted to avoid. Marcel moaned: "All the passion, the ferocious egotism comes from Mrs. Curie, and that Langevin, being weak, gives in to a stronger will who cannot even so totally detach him from his children . . . And what a disgrace! To have the name of Curie and not to be able to wear it nobly, and to keep proper to his daughters. How very clear things become now! For months, Langevin's winnowed, exhausted, wrinkled appearance, which, had it been for any other, would not have left any hesitation about the sort of fatigue!"[15]

Again, Brillouin could only see Pierre Curie on a pedestal, the revered scientist, whose wife was making a mockery of his name by pursuing a chance at love with a married man, a man who had no choice in the matter, but had given into her seductive charms and who sapped his very strength in the process. It seemed to Brillouin that Langevin's children were the only thing that kept Paul from jumping headlong into the abyss of this tawdry affair. Even so, Marcel appeared blind to the fact that there were two people, both with free will, who were participating in this drama, not allowing Marcel to view things from the two different perspectives of each protagonist. Brillouin's conservative cultural viewpoint could not provide for anything approaching an equal status of women and their rights in early 20th-century France, where most still clung to the notion that a married man was entitled to his indiscretions, as long as they weren't paraded in public.

Two days later, Brillouin's journal entry started with a message from Henri Bourgeois that even though Bourgeois offered to return the love letters to Curie if she ended the relationship with Langevin, she refused. Brillouin decided to find Jean Perrin in his laboratory at the Sorbonne to discuss the whole situation, but bumped into another professor, Paul Appell, who was a longtime friend of the Curies. The entry continued: "[Appell] gets around to asking for news of Langevin, and to talk about his household, Mrs. Langevin's bad character, and domestic arguments—so I cease to wonder if I should remain silent . . . I let [Appell] know sufficiently of the threat hanging over the University . . . he suggests that [Curie] be advised to take some rest and to leave [on vacation]—(without Langevin!)"[16]

Brillouin clearly saw that the rumors of the Curie/Langevin affair were spreading to a wider circle of people within the university system, Paul Appell being one whom Brillouin might have thought was unaware of the situation until this unplanned meeting. Marcel then proceeded to contact Jean Perrin with his concerns about Curie and her situation, his relatively short note reading in part: "I did not ask you to come and talk about [the Curie/Langevin affair] with me because I feared that I would express my indignation with too much vivacity to be of any use . . . The dead are quite dead and do not know the future. It is important for the honor of all that those living who are aware of the present remain as few as possible."[17]

A letter reply quickly came from Perrin that was friendly, logical, and very much to the point. Perrin admired Brillouin as one of his former professors and expressed respect for him as a friend and colleague. In the hope that they could see the issue similarly, Jean made an analogy concerning the Dreyfus affair, still on many people's minds in France. It was an incident that had torn the country apart many years before over the unjust jailing of a Jewish captain in the French military who was falsely accused of being a German spy.

The trial had created "Painful divisions . . . between families and friends who until this time had been totally united. But at that time, [you and I] judged in the same way and, with all our force, we fought so that in under no circumstances could a man be condemned without being able to defend himself. It will certainly suffice for me to ask you only once to act today as then. You pronounce a condemnation that is severe and certain. And yet, out of three people involved, only one gave you information . . . this one-sided information exposes you to the most serious of mistakes . . . Finally, I would see it as a weakness on my part, considering all that I know up to this day, if I did not tell you that I find Mrs. Curie to be one of the most noble beings that I have ever met. Please see in these words a peaceful affirmation that in no ways gives hope of imposing itself on you."[18]

Perrin appealed to Brillouin's prior actions during the Dreyfus affair, when both Marcel and Jean had stood firm on the rights of an individual to defend oneself, and tried to convince Marcel that the same principle applied here as well. Brillouin had heard Jeanne Langevin's side of the story, but without receiving input from Paul and Marie, how could Marcel determine the real situation? Beyond that, Perrin insisted that Marie Curie's character was beyond reproach.

Shortly after this exchange of letters, on April 26 Perrin visited Brillouin at the College of France, and a long journal entry followed: "Perrin claims that Mrs. Curie is not in any way pushing for divorce—that there is no sexual relationship between them. I observe to him that supposing that this is true, it is completely incomprehensible that having three labs in which they can work . . .

they should need to rent a pied-a-terre under a false identity—unless it is for activities other than work. Proving the intellectual nature of their meetings at this pied-à-terre would seem to me impossible to do.

"Langevin seems to complain a lot about his unhappy marriage—his jealous, violent, and foul-mouthed wife. A supposed bicycle accident, at Palaiseau, in the first years of his marriage, which had left him disfigured, was due to a garden chair that Mrs. Langevin had thrown at her husband's head, etc., etc., constant accusations about Langevin's mother and her lovers. Criticism for the excessive spending of his father and brother, etc.—More recently, to their son (9–10 years old) 'If you were to take a mistress, you would choose a young and pretty one, etc.' In one word a fury."[19]

Brillouin and Perrin discussed the affair, with Perrin taking the position that it was not sexual, Brillouin incredulous that Perrin could not see what was plainly evident from the fact that Langevin and Curie had established a secret meeting place, under an assumed name, to carry on their relationship. Perrin described Langevin's continuing issues with his wife, including the well-known incident of her attacking him with a chair, and ending with Jeanne's advice to their young son to take a mistress who was "young and pretty," implying that Curie was neither.

Perrin then told Brillouin that Langevin had actually been living at Perrin's place for a few weeks, and wouldn't go home unless the love letters were placed in the hands of his lawyer. Brillouin wrote of these letters, "Perrin says that he has read them and can find nothing more than a very warm affection."[20] In Perrin's view, they were not quite the powder keg that was ascribed to them by everyone else who had taken a look. Then, Brillouin stated how he envisioned this to all play out: "I do not want to get mixed up in all of this; it's just a matter of time of avoiding a first scandal which would be an act of violence by Mrs. Langevin on Mrs. Curie, which is not at all impossible given her present manifest state of exasperation—and a second act of violence which would be the publicity given to the cause of this act of violence, and to the letters in question—with the proven interpretation that the public will make of them and which is not very flattering for Mrs. Curie and [Langevin]—(and that for my part I look at as fair)."[21]

Brillouin clearly felt that the situation was much more volatile than Perrin imagined. In fact, he predicted violent reactions, the first of which was one between Langevin's wife and Curie, which had already been threatened one evening the previous year when Jeanne accosted Curie on the street and demanded Marie leave town or Jeanne would kill her.[22] The second action, publicity of the affair and the love letters, would take place as well, in outsized newspaper headline form during November 1911, an exposé of the relationship between Curie and Langevin that was embarrassing for the two but, in Brillouin's own observation, probably deserved and "fair."

Jeanne Langevin came to visit Brillouin's wife Charlotte on yet another occasion, in early May, providing the last major entry in the journal Marcel had kept on the activities surrounding the Curie/Langevin affair. There were still some important details that Jeanne wanted to provide: "Mrs. [Langevin] comes back to see my wife—and to insist on the fact that when her husband was ill, with his keys, she had gone to the pied-à-terre—where she had taken the letters—where there is a bed and objects necessary for female intimate hygiene which can leave no doubt—my wife had quite some trouble in getting her to leave and in stopping her by telling her that I am unable to do anything about it and that this is none of our business . . ."

"Mrs. Langevin came back again (July) bringing a copy of Mrs. Curie's letter; wanting to have a woman reread it, and leave it with [Charlotte]. My wife has great difficulty in preventing her from reading it and in making her take it away with her—repeating that we can do nothing about it, and that a copy of a letter doesn't prove anything and that it is not our business—My wife is fed up with all this messy gossip . . . Let them get on with it, them and their friends who took sides in one way or the another. The water seems murky enough to me. Let it not harm any of mine or me myself, nor the University or the College—or result in unhappiness for the Langevin's children!—but I have no means to remedy it. Unless something unexpected happens—matter closed on the July 21 1911."[23]

Jeanne Langevin wanted to prove to the Brillouins once and for all that this illicit affair had gone too far. She especially stated that it was she who, upon taking her husband's keys while he was ill, had gone to the apartment Paul and Marie had rented and removed the love letters she found there. In her inspection of the apartment, she had come across a bed and items of feminine hygiene that were conclusive proof of the sexual encounters taking place in the room.

Numerous accounts of the Curie/Langevin affair have speculated about who actually had stolen the love letters that were to be the "smoking gun" pointing to the reality of the relationship. Some say Mrs. Langevin hired a detective to break into the pied-à-terre and take the letters.[24] Another deemed the thievery as merely semantics, the simple fact being that, whoever took them, the letters ended up in the hands of Jeanne Langevin and her brother-in-law, to be used later to smear Langevin and Curie.[25] A few discuss the sexual nature of the affair as conjecture rather than reality, pointing out that the letters themselves can be read to have shadings of meaning that might make conclusive arguments about the issue difficult. The riddle of the mysterious thief of the Curie/Langevin love letters had a few potential solutions, but the simplest, expounded upon by Brillouin in his journal, was that the jealous wife of a cheating husband went to his rented rooms in Paris and took them herself to prove the existence of an affair upon which everyone had differing opinions.

Brillouin's writings certainly provide some insight into how those on the periphery of the Curie/Langevin affair were nevertheless drawn into its gathering storm. They also shed light on the nature of a conservative Frenchman's reaction to these activities, in this case Marcel being quite judgmental concerning Curie and worried as much about his own reputation and that of his academic institution as with the individuals involved.

In early 1912, after the publicity of the affair had spread around the world, Brillouin received a letter from the venerated facilitator at the First Solvay Conference, Hendrik Lorentz. The Dutch physicist, in conjunction with Ernest Solvay, had taken the discussions of the Solvay Conference to the next level. He would serve Solvay as organizer and eventually chairman of a committee, mainly composed of attendees to the conference, being assembled to establish, organize, and administer an International Institute of Physics, supported once again by Solvay's largesse. In a communication of late January, Lorentz explained his activities to Brillouin concerning the creation of the proposed institute and its commission to move the organization forward. In doing so, it seems Lorentz thought highly enough of Brillouin to seek his opinion on the status of the Curie/Langevin affair. Specifically, he sought advice on how to potentially proceed with Marie Curie's participation in Institute affairs: "I read the details of this sad tragedy and I wondered if these facts should have an effect on Mrs. Curie's position in the New International Scientific Council [of the Institute]. Well, I think that the position must not be affected by it and that we still adopt the same attitude toward her as that which we adopted in Brussels when all of this was unknown to us. It also seems to me that I should not consult Mr. Solvay on this point nor the other members of our commission. However, I would appreciate it very much if you could give me your opinion and I would ask you to tell me if you share this opinion, and if, like me you would not find it suitable to bring this up with Mr. Solvay and our colleagues."[26]

It's obvious that Lorentz was concerned about the affair, but felt the participants in both the conference and the resulting proposed Institute committee should look past this unfortunate issue rather than judge Curie harshly professionally for the personal situation in which she was embroiled. Lorentz was suggesting that there was no need to even bring this up at all to Solvay himself, but rather to settle this at the working level of the himself and the committee.

It appears Brillouin composed a draft letter of response in late January or early February. A cramped, four-page, unsigned missive was constructed with Brillouin's printed trademark journal heading of "College de France, Physique Mathématique" in the upper left-hand corner of the first page. The writing was so tightly packed that it completely filled each page of paper, afterthoughts captured in notes jotted on the margins of the pages themselves as more information poured forth from Brillouin's pen. Langevin's friend felt he had something important to say and seems to have kept thinking of more to add as he pondered what he had already written. Brillouin undoubtedly thought this was an opportunity not only to share his own feelings on the matter, but to be something of a representative of a proud French scientific community to the Dutchman Lorentz on an issue that had shocked or outraged many.

Brillouin began his draft under the heading "confidential" and stated this again with sincerity in an addition (noted here) that filled the whole left margin of the opening page of the letter and covered the top of it as well, actually obscuring the written date: "You are, Sir, by your character and your nationality, the only foreigner in whose esteem, sympathy, and righteousness I have confidence enough to write to you about my sentiment on this most painful subject without fearing to be poorly judged myself. I take responsibility for what I am writing as a Frenchman, but it would be painful for me if my feeling gained international publicity. I thus desire that this letter remain confidential."[27]

Brillouin proceeded to take Lorentz back in history to recount details that had brought everyone to this shameful moment. He described Langevin as a student exhibiting the promise of perhaps being "destined to become the number 1 physicist in France . . . during the first years of the marriage, Mr. Langevin was devotedly affectionate, Mme Langevin an irreproachable mother and wife . . . I have been told now that from that time on, there had been violent arguments between them and vulgar language on the part of Mrs. Langevin, and from that time on Langevin was unhappy . . . that the arguments should become more violent, perhaps vulgar; I am told now; it is possible, I have no idea; but I have no indication that at the bottom of things, it should be she who is in the wrong . . ."[28]

Brillouin then commented on Marie Curie not being French in origin. If she had been, he felt the issue would have been quickly ruled in Mrs. Langevin's favor. He noted that "Those who are familiar with the Slavic world tell us that the Slavic mentality is impenetrable, and you can't conclude as if it was a Frenchwoman . . . It is because she is of an intellectual value of the highest order that Mrs. Curie had so much power over Mr. Langevin. And yet, we could not be forgiven for thinking that [Pierre] Curie—whose memory was worth more—and Langevin are possessed of quite superior minds, the former with more activity and a stronger and more noble character, the latter possessed of a wider and more penetrating culture from a mathematical point of view."[29] To Brillouin's way of thinking, for a variety of reasons, Marie Curie ranked behind her husband Pierre and her lover Langevin intellectually as well as culturally, especially since Marie wasn't even French to begin with, but of Slavic background.

Brillouin finally arrived at the point of the letter, giving to Lorentz his advice on Marie Curie's worthiness to sit on the committee Lorentz was assembling for the new Institute:

> What to conclude?
>
> 1. Mrs. Curie represents the discovery of radium, and the in-depth studies of its properties; a colossal work superior

to that of most physicians [physicists]of this century, with inestimable consequences.

2. She does not represent the French physicians. Very few would take her as their intermediary for their requests at the Solvay Foundation—especially among those who study radium outside of her laboratory and do not belong to the ardent chapel in her entourage.[30]

It had taken Brillouin almost four pages of detailed prologue and justification for him to arrive at the declaration that Marie Curie was not French, not one of *us*. As such, how could she continue to represent the French world of physics on the Solvay Institute committee? Curiously, as a postscript jammed into the margins of the letter, Brillouin was quick to offer a disclaimer on his own past personal contact with Marie Curie, "To develop my impressions, you should also know that in reality, I do not know Mrs. Curie. I have never had more than five minutes of conversation with her during my life. I only went into her lab once (in 1904–05) to see experiments . . ."[31]

We assume it was this draft, or something close to it, that was finally sent to Lorentz. As everyone knew, Hendrik Lorentz was as compassionate as he was wise. His comments back to Brillouin on the matter were contained in a short paragraph at the end of a lengthy letter of response, mostly on more general issues related to the organizing of the Solvay Institute and its leadership committee roles and responsibilities. Lorentz was courting Brillouin as a potential committee member for the Institute, and so needed to treat him with more than a modicum of respect and deference. Nevertheless, the Dutchman's decision concerning Marie Curie's involvement could only add to his reputation: "My wish that the International Board be composed of all the present members thus includes Madame Curie. Although your letter has given me the impression that she carries a serious responsibility, it is not up to me to judge her. I believe that I do not have the right to carry out any act which might lead to her exclusion from the board in which she could occupy a place considering her eminent talents and for her services rendered to science . . . I will not fail to consider your letter as strictly confidential."[32]

In Marie Curie's pursuit of her impossible dream, the existence of love letters between herself and Paul Langevin had proven sensational. But they were a mere sideshow to the very human issues at hand in turn-of-the-century France that her personal affairs had come to embody, writ large: the equal rights of women, specifically in this case the impossibility of a woman to have the same right to pursue a relationship with a married man as that same married man had to pursue a relationship with a woman who was not his wife. French culture could not tolerate a mistress overriding the virtues of a wife, even an abusive wife. It was too large an obstacle to overcome, especially in this case if that mistress was the Polish foreigner Marie Curie. Some in Europe could forgive and even look past the personal complexities of this situation, focusing on the professional contributions of the individuals, while others could only observe and respond with criticism and rejection. Unfortunately, a number of those on the Nobel Prize Committees on Physics and Chemistry could be classed in this second category, which would be seen in their reaction to the publicity that was to shortly accompany the Curie/Langevin affair.

CHAPTER NINE

Action and Reaction

The largest recorded explosion in history took place on August 27, 1883. It was the equivalent of detonating approximately 200,000,000 tons of TNT, 10,000 times the force of the atomic bomb dropped on Hiroshima, Japan, at the end of World War II. The devastation caused by the blast had worldwide implications, both immediate and far-reaching. This was the impact of the eruption of Krakatoa, a volcano on an uninhabited island in Indonesia, between Sumatra and Java in the South Pacific.

The volcano had been making itself known since ancient times. Periodic explosions had been recorded intermittently for more than thirteen centuries prior to the 1883 catastrophe. Earlier in the year, the volcano had been exhibiting ominous signs of activity for months. Then at ten in the morning on that fateful day in August, Krakatoa unleashed all of its awesome power. A cacophonous eruption could be heard almost 3,000 miles away in Australia, the volcano spewing molten lava, pumice, and chemical dioxides of sulfur and carbon into the air. The tremendous blast caused most of the island to literally disappear as the volcano itself collapsed into the ocean. Tsunamis formed, generating some waves as tall as the Statue of Liberty. These watery walls of devastation slammed into more than 160 towns on the islands of Sumatra and Java, killing over 36,000 people.[1] Only the lack of major concentrated population centers in the vicinity prevented the human toll in the immediate aftermath from being far greater.

As devastating as the actual explosion was, the remnants of the eruption were to have broader implications. The particulate ash and sulfur dioxide produced in the event were sent high into the stratosphere, eventually carried around the world by the winds. The ash, as minute dust particles, acted to scatter blue wavelengths of sunlight, leaving the mix of remaining colors of light to produce evenings filled with vivid magenta sunsets around the world. The sulfur compounds created a more significant effect on global temperature over the next few years. The sulfur dioxide mingled with water in the atmosphere and was converted to sulfuric acid, which formed sulfur-based aerosol droplets that reflected sunlight back into space rather than letting it pass through to Earth, consequently having a cooling effect on the world's temperature.[2] Scientists observed that there was an overall average decrease in climate readings of about 2.2 degrees Fahrenheit.[3]

One of the many scientists interested in this type of cataclysmic volcanic phenomena as it related to climate was the Swedish university professor Svante Arrhenius. He had previously conducted extensive work in other fields, discovering the electrical nature of a class of chemicals called electrolytes that formed positively and negatively charged particles, known as ions, when dissolved in water, explaining electrical conductivity of solutions. This approach was so controversial when he first proposed it in a crude form as part of his doctoral dissertation at Uppsala University in Sweden in 1884 that it was ill-received by the committee reviewing his efforts.[4] He ultimately won over the world with refinements to his theory, but during the 1890s his interests, like those of a number of other scientists, shifted to atmospheric phenomena that included investigating the cause of glaciers formed during the last ice age.

Arrhenius was especially intrigued by the role carbon dioxide played in relation to the temperature of Earth's atmosphere. It was known that volcanic eruptions belched significant quantities of both sulfur dioxide and carbon dioxide into the air. Sulfur dioxide ultimately was found to cause atmospheric cooling, manifested in the previous decade by the literal chilling effects of the Krakatoa explosion that lasted for a number of years. Decades before this, in mid-century, had come isolated experimentation, chiefly by the Irish physicist John Tyndall, establishing that carbon dioxide acted to inhibit Earth's release of heat back through the atmosphere into space.[5] An amateur American scientist, a woman from New York named Eunice Foote, through detailed investigation had reached a similar conclusion a few years earlier. Her independent study of the issue appears to have been generally bypassed in the scientific community, certainly not unusual for women scientists of that era.[6] Both felt the phenomenon they had studied could be thought to ever so incrementally warm the planet through additions of carbon dioxide into the air, which periodically occurred through volcanic eruptions. Arrhenius applied this knowledge to determine what would happen if the levels of carbon dioxide in the atmosphere were lowered or raised on a consistent basis rather than just intermittently through natural volcanic activity.

To this end, Arrhenius employed reams of scientific measurements that had recently been taken of carbon dioxide levels in different parts of the world to laboriously construct a chart that showed average changes in global temperature as a result of variations of carbon dioxide in the atmosphere. Even though Foote and Tyndall had shown that water vapor caused a similar heat-trapping effect as carbon dioxide, Arrhenius focused on the latter gas because water vapor levels varied on a daily basis as opposed to the levels of carbon dioxide, which were relatively permanent additions to the atmosphere once embedded by volcanic eruptions.[7] His calculations showed that reducing the carbon dioxide level in the atmosphere by 50 percent would lower European temperatures by four to five degrees Celsius (about seven to nine degrees Fahrenheit), moving the world toward ice age–type temperatures.[8] But, he also knew that the relatively recent industrialization of the world was utilizing coal to fuel it, a more steady source emitting carbon dioxide into the air than the sporadic addition from volcanoes. Arrhenius calculated that if coal was burned at its current rate of use of the mid-1890s, it would take approximately thirty centuries for the carbon dioxide emissions generated by this activity to double and cause Earth's temperature to heat up by an average of five to six degrees Celsius (nine to ten degrees Fahrenheit).[9]

Of course, no one could have anticipated that the emission rate of carbon dioxide into the atmosphere would accelerate even more dramatically from end-of-19th-century levels as worldwide industrialization, with its burning of coal, then oil and its derivatives, relentlessly increased during the 20th century, ultimately raising carbon dioxide levels in the atmosphere at an exponential rate. But the early work of Foote and Tyndall, combined with the calculations of Arrhenius, had laid the scientific basis for understanding how Earth's temperature could rapidly accelerate, caused by industrialization supported by the burning of fossil fuels.

After completing his climate work, as the 19th century opened Arrhenius continued to examine other scientific puzzles on the macro level of Earth, the planets, and the universe. At the same time, he was intrigued as well by the chemistry of immunology. Although not as successful in these efforts as with electrolyte chemistry and climate investigation, he had proven to be a dedicated, insightful scientist with eclectic interests. In addition to his various investigations, he had also become an active participant in the review of candidates and final selection of the Nobel Prize winners in physics and chemistry. This important initiative had been created in the will of millionaire industrialist Alfred Nobel upon his death, its execution invested in the Royal Swedish Academy. Through his dedication to this project, as well as his networking within the Swedish academic community, some might argue that Arrhenius was the singular most powerful voice among the Swedish professors and scientists involved in the complex task of arriving at worthy awardees. He himself had won the prize in chemistry in 1903 for his ionic conductivity work with electrolytes. This, despite the initial poor reception that his doctoral thesis dissertation on the subject had experienced years earlier, which he felt had almost cost him his scientific career before it had even begun. Now adorned with a Nobel Prize of his own, Arrhenius had become fiercely protective of the professional stature that he felt accompanied the award.

The same year that Arrhenius won his Nobel in chemistry the Curies shared the physics prize with Henri Becquerel for the discovery of radioactivity. He had first met the Curies in Paris at the international physics gathering there in 1900, had a high opinion of them, and was one of the strong proponents of Marie Curie being nominated for, and eventually winning, the Nobel Prize in chemistry in 1911 for the discovery of polonium and radium and the isolation of the latter. However, his determination to keep the awards as free of controversy as possible led him to at times be almost paternalistically overprotective. For years he had waged a private war against a professional rival, Walther Nernst, in denying him a Nobel Prize. This was in large part due to his puritanical view that Nernst had compromised his scientific credentials and debased himself by mixing business with science with his very profitable sale of patents related to his namesake lighting invention, the "Nernst lamp."[10] Now, he brought his self-righteous moral code to bear on Marie Curie and her award of the 1911 prize, claiming that if he had known of the intense publicity her affair with Paul Langevin would engender before the award was announced, her award nomination might have taken a different turn.

The initial newspaper article on the Curie/Langevin affair broke on Saturday, November 4. The Parisian newspaper *Le Journal* published an article strewn with innuendo, including the potential discovery of love letters between the pair, but light on facts. The next day *Le Petit Journal*, where Jeanne Langevin's brother-in-law Henri Bourgeois was an editor, published a similar piece describing Jeanne's marital suffering, both physical as well as emotional, caused by a husband deserting her for the home-wrecker Marie Curie. To its credit, *Le Petit Journal* subsequently printed a spirited response from Paul Langevin, who said his marriage had long been troubled and that he had left his wife months ago, proceeding to counter with his version of his wife's physical abuses toward *him*. Marie's own initial response came through *Le Temps*, a newspaper that was sympathetic to her plight. In her controlled fashion, she stated that the article in *Le Journal* was laden with falsehoods, complained about the injustice of the whole situation, and then very pointedly threatened lawsuits against any further improper reporting. The threat of legal action was scarcely cause for concern in 1911 Paris with its lax libel laws. Nevertheless, an apology was quickly forthcoming from the original offending journalist at *Le Journal*, Fernand Hauser, who almost claimed temporary insanity as he questioned his own professionalism in writing such an article.[11] Arrhenius tried to keep abreast of all the various reports, and there were many that followed.

The proliferation of daily newspapers and journals in Paris in the early 20th century was almost overwhelming, more papers providing more points of view than almost imaginable. As well, the last two decades of the previous century had witnessed the liberalization of French government regulation of the press, allowing increasing excesses that created sensationalism as a journalistic standard.[12] *Le Journal*, along with *Le Petit Journal* and *Le Matin*, were the major players in total circulation and gossipy reporting, while *Le Temps* was another large paper known more for accuracy of the news.[13] Half a hundred others, of every political and philosophical stripe, rounded out the daily offerings in the city, catering specifically to those who wanted to read a paper that aligned with their own views in conveying the events of the day.

While the actions and reactions of the journalists, the Langevins, and Curie were beginning to transpire in the press, on Tuesday, November 7, the first public announcement of Marie Curie's winning an unprecedented second Nobel Prize, this time in chemistry to bookend her initial prize in physics, was notified to the world by Reuters news service. Curie was now not only the first person to win a second Nobel Prize, she was also the first person to win the prize in two separate disciplines. And, of course, she was the only woman to ever have succeeded in winning a Nobel Prize of any sort, now having done it twice. By all rights, France should have been bursting with pride over these amazing scientific accomplishments from one of their own, a woman who had already become famous around the world for her professional achievements. But the preceding exposé of Curie's affair with Langevin only days earlier had changed all that. Even the sympathetic *Le Temps* waited a few days after the Nobel information had initially been provided by Reuters to print the news of Curie's award. On November 9, they noted this groundbreaking feat, not displayed in bold print on the front page but found well within, almost as an aside on page four.[14] Most papers covered the Nobel award similarly, buried deep under the avalanche of headlines starting each day with mounting outrage over the affair.

Indeed, the liaison was juicy with details that could be presented in many ways. Most editors chose the low road, spinning a tale of a woman who had compromised a married man five years her junior. A husband with a long-suffering wife, a scientist who had been the protégé of his mistress's husband before his death. Not to mention that the married man had a family with four children. Even more tantalizing was the fact that the conniving temptress was Marie Curie, wife of the admired, deceased Paul Curie, arguably one of the most well-known scientists in France, if not the world. So Marie obviously had no shame as well as no respect for her dead husband. To top it all off, Marie Curie was not French at all, but an immigrant from Poland and suspected of being Jewish. With nationalism more proudly espoused than ever at this time, being Polish didn't help. The French relationship with Germany, never good to begin with, was continually deteriorating in a buildup to the coming world war, with those not of French descent always suspect. The Dreyfus spying incident of more than a decade earlier was still an open wound for French society's view of itself versus the dreaded Germans, and anti-Dreyfusards called into question the Curies' pro-Dreyfus past leanings. As always, being labeled a Jew was still a stigma that was easily applied when a smear campaign called for it, whether true or not. Finally, and inevitably in that era, much of the issue came down to Curie being a woman who simply did not know her place. The smoke had not totally cleared from the beginning of the year, when Marie Curie had dared to promote herself as a worthy candidate in her failed bid for admission to be a member of the all-male French Academy of Sciences. Now, she was attempting to steal away a prominent Frenchman from his rightful wife, no matter his own feelings on the matter or how dismal their marital relationship had become.

The smaller newspapers and journals now proceeded to jump into the fray to deliver to their narrow slice of the Parisian public their unique social perspectives on the steamy scandal. *L'Action Française*, run by the well-known reactionary, anti-Semitic, and outspoken anti-Dreyfusard Leon Daudet, knew exactly how to appeal to its audience. It presented the situation as one tailor-made for their brand of scathing ultraconservative, Francophile-to-the-core skewering of a manipulative foreigner, a misguided, self-seeking woman. A murky conspiracy theory emerged from the pages of Daudet's paper, a plan led by a combination of high-placed, progressive government officials and influential media moguls. Daudet insisted that this group was intent on preventing the French public from seeing Marie Curie for the evil force in society that she represented, one of degrading liberal moral standards that traced much if not all of her attitudes and actions to her prior pro-Dreyfus support.[15] Curie had indeed employed her influence on some of her newspaper connections to try to stem the mounting tide of published stories of ranging veracity concerning the affair, but the storm abated for only a short time.

As the rumors of the existence of the Curie/Langevin love letters escalated, Gustave Téry, the editor of the journal *L'Œuvre*, another ultraconservative, culturally prejudiced weekly newspaper, obtained and published extracts of these letters that were available as part of the actual text of a separation proceeding that Jeanne Langevin had instituted against her husband, to be heard in court on December 8.[16] The letters from Marie were touchingly affectionate in parts, while sternly directive in others, which were the carefully selected portions that appeared in print. In these, among other guidelines, she cautioned Langevin not to sleep with his wife if she tried to win him back via the bedroom. She told Paul that if he was to get his wife pregnant, "it would mean a definite separation between us . . . I can risk my life and my position for you, but I could not accept this dishonor."[17] Her instruction was understandable under the circumstances, Curie desperate to try to convince him to move on together and leave his wife behind. But it came across as manipulative, and was even used by the conservative press to show that Curie wanted to deprive France of her rightful progeny by keeping a lawful wife from getting pregnant by her own husband. Marie Curie was again plainly overstepping her role as a woman, the right-wing press outraged at her attempts to apply reasoned scientific thought in what they felt was an unscrupulous maneuvering to separate Langevin from his wife.[18]

On November 23, the day Téry's paper published these carefully edited portions of the letters found in Jeanne Langevin's public legal complaint against her husband, Albert Einstein composed and sent Marie Curie a letter of his own. Einstein had established a budding friendship with Curie earlier that month as they exchanged views on various aspects of physics, the energy quantum, and radioactivity at the First Solvay Conference. The French coterie at the gathering had taken a definite liking to Einstein, especially Curie and Langevin, the latter having previously invited him to lecture in Paris. Likewise, Einstein found the French interesting and far from standoffish, especially Marie Curie. Their minds seemed to be attuned to one another, and after seeing the continuous bombardment of Curie in the press throughout the month, Einstein felt it was now time to reconnect in a manner suited to a true friendship. In his note from Prague he offered sincere support: "I am so enraged by the base manner in which the public is presently daring to concern itself with you . . . I am compelled to tell you how much I have come to admire your intellect, your drive, and your honesty, and that I consider myself lucky to have made your personal acquaintance in Brussels . . . If the rabble continues to occupy itself with you, then simply don't read that hogwash, but rather leave it to the reptile for whom it has been fabricated."[19]

Curie, a dozen years Einstein's senior and infinitely more experienced in the public's ever-changing loyalties, still was no doubt grateful for the sentiments expressed in the letter by this newly found and respected comrade. Others within the French scientific community had offered similar input, including fellow French conference attendees Poincaré and Perrin as well as her deceased husband's brother, Jacques. But it was Einstein, from outside the tight-knit French circle of brilliance, of Germanic origin at that, who offered such an unexpectedly direct, sympathetic message of concern and solidarity.

Einstein signed off with regards to Paul Langevin as well, someone who had had enough of the newspapers providing the public with a feeding frenzy on misrepresentations of Curie's character. Two days later, on November 25, Paul Langevin and Gustave Téry met for a pistol duel on their chosen field of combat, a velodrome bicycle stadium in Paris that had frequently served as the venue for these types of contests. Langevin had felt he had no choice but to issue his dueling challenge to Téry, who had called him "a boor and a coward" in one of his articles concerning the affair, words that leapt off the page at Langevin as a searing accusation requiring defense of his honor.[20] Curiously, it was Marie Curie on which the brunt of the article's attack was focused, yet Langevin's decision to fight seemed to be arrived at based on his pride being bruised by Téry's superficial name-calling, rather than in a spirited defense of his lover's sullied reputation.

By the beginning of the 1900s the ancient custom of dueling, which had been slowly dying out as the world became more civilized, actually proliferated in many countries on the European Continent. Hundreds were fought each year with swords or occasionally pistols throughout Italy, Germany, Austria-Hungary, and France. Contrasted with the ruthlessness of Germanic contests, duels in France were much more tame. There, the very act of drawing a bit of blood from one of the contestants was more than enough to signify the willingness of the duelists to symbolically risk their lives for their sacred honor. Often, a duel in France was thought comical by others for its comparative lack of physical danger. Among those scoffing at the French version of a duel was Mark Twain, who, in comparing the Austrian duelists to those in France, described, "Here he fights with pistol or sabre, in France with a hairpin—a blunt one. Here the desperately wounded man tries to walk to the hospital; [in France] they paint the scratch so they can find it again, lay the sufferer on a stretcher, and conduct him off the field with a band of music."[21] Yet, duels in France, as across the Continent, accompanied the building nationalistic tensions in the years before the coming world war. They were an intense display of masculine pride, a mixture of fearlessness and fealty to somewhat dated principles of honor and chivalry accompanied by the acceptance of potential danger in an ever more xenophobic world.

This was actually one of at least five duels that were the direct result of the competing thrusts and parries of razor-sharp charges and determined defenses posited on the Curie/Langevin affair by the newspapers. Just the day before, Téry had fought a sword skirmish with Pierre Mortier, an editor of the literary journal *Gil Blas* and a friend of Marie's, who *had* fought in defense of her honor, not his own. Mortier paid for his friendship with two wounds, a minor arm cut and a more serious stab to the wrist, stopping the encounter.[22] It was all captured in a brief movie moment, a cameraman present along with numerous photographers as part of the sensationalism of the day coupled with the popularity of the new medium of film.[23] It seems that the periodical *Gil Blas* had a number of journalists sympathetic to Marie Curie's cause. Henri Chervet, another editor at *Gil Blas*, had just fought a similar sword duel with *L'Action Française* editor Daudet the day before the Mortier/Téry contest. Chervet bested Daudet in this encounter, giving him a small arm wound, likewise caught on camera to be played in daily newsreels shown to the curious masses in local movie-houses.[24]

At least two more rounds of swordplay were to take place after the Langevin/Téry pistol duel, one on December 23 between Mortier and an associate of Leon Daudet's at *L'Action Française*, Jacques Bainville. Then Mortier participated in yet another duel, this one the last in the series, when on February 24, 1912, he faced the irrepressible Leon Daudet, who had previously lost to Mortier's colleague Chervet in

November. This time, Daudet won the contest by wounding Mortier, who seemingly had been willing to stand up to the tirade of anti–Marie Curie sentiment by freely shedding blood, albeit in minute doses, to protect Curie's honor by participating in three duels on her behalf in three months.

Certainly, at least from this sequence of encounters, it appears that French journalists were ready to defend their honor, as well as that of others in their esteem, at the drop of a pen. With the relative loosening of libel laws for the French press toward the end of the 19th century came the increasing proclivity of those writing ever more sensational prose to stand proudly, if not recklessly, behind their work. Not only editorials, but daily articles, were freely signed for everyone to associate newspaper stories with specific names so that, as one reporter and noted duelist put it, "Behind every [French] journalist everyone expects to find a chest."[25] When editor Téry blatantly insulted Curie and Langevin, the latter felt he had to stick his chest out as well and answer with a sword or pistol. Téry then proceeded to describe Curie as "one of those women who wanted to enter male citadels [like the French Academy of Sciences] if she could but fall back on 'French gallantry' when obstacles arose."[26] Langevin felt he had no choice now but to go forward with the duel.

Perhaps a bicycle race once around the long oval velodrome track would have been more of a contest, at least providing a clear winner. For, when Langevin and Téry were faced with raising pistols for one shot at each other from twenty-five meters, neither had the desire to pull the trigger. Langevin's second, fellow College of France professor and future French prime minister Paul Painlevé, was as unfamiliar with the finer points of pistol dueling as Langevin himself. It was fitting that, as a mathematics instructor, he was charged with the duel countdown. But ironically, when it came time to perform this simple duty, his inexperience showed as he rushed the count such that Téry wasn't ready to take aim when Langevin already had his pistol up and in firing position. Téry never raised his gun, and Langevin slowly lowered his own. Each party's second looked at the other in embarrassed silence and, after a nervous exchange of words, took the weapons from the opponents and fired them harmlessly into the air, ending the confrontation. The newspaper *Le Petit Journal* reported that indeed no shots were fired at the participants, both of whom claimed that they respected the other too much to kill their opponent.[27] The relative non-event was covered to varying degrees across Europe and the world, but a particular shockwave from the confrontation was felt in Stockholm.

Arrhenius had been nonplussed by it all. He and Marie Curie had been exchanging letters throughout the month as the publicity of the affair was accelerating. The first blast of innuendo, then accusation, had subsided, but the reporting gathered momentum again and by mid-month articles on the scandal were found in print everywhere across Europe, culminating in the news of Langevin's duel with Téry on November 25. Arrhenius was nervously following the situation, monitoring each headline and trying to decide what action he felt he, as self-assumed guardian of the honor of the Nobel Prizes, should take to potentially influence Curie's actions and protect the reputation of the awards. As Elisabeth Crawford wrote in her book on the early Nobel Prizes, Arrhenius had a "real fear of adverse publicity [which] lay behind his attempt to prevent Mme. Curie from traveling to Stockholm to receive her second Nobel Prize in the midst of the uproar in the press over her presumed liaison with P. Langevin."[28]

Initially, Arrhenius felt that the criticism of the pair was unjust, basing his feelings on the idea that this was a smear campaign against the two being conducted to stir up sales for the French newspapers. However, as the month wore on, the articles intensified in their accusations against them, culminating in the publishing of a portion of the stolen love letter correspondence by editor Gustave Téry of *L'Œuvre*.

He was intent on dragging both Curie and Langevin through the mud in a public display of sympathy to Paul Langevin's wife, whom Téry painted as a paragon of French family virtue. Finally, in the last week of the month, news of Langevin's duel tipped the scales against Curie and Langevin, at least in the mind of Arrhenius. He could wait no longer, sending a letter that essentially demanded Curie stay away from the Nobel Prize award ceremony on December 10 in Stockholm. The letter Arrhenius wrote to Marie Curie on December 1 read in part: "I asked a few colleagues what they believed should be done in the situation, which was actually made worse by Mr. Langevin's ridiculous duel . . . All my colleagues told me that it is to be hoped that you do not come here on December 10 . . . Honor, esteem for our Academy as well as for science and for your homeland seems to me to require that in such circumstances that you desist from coming here to take the prize."[29]

Curie's previous correspondence with Arrhenius during the last few weeks of November had led her to believe that he understood the situation to be one that was being vastly overblown by the French press. Now, Arrhenius's change of attitude expressing to Curie that she abandon her acceptance of the award, at least for the time being, was unexpected and disheartening. Simultaneous to this letter exchange, multiple telegrams from another within the Swedish academic community had been sent to close colleagues of Curie asking that she come to Stockholm to receive her prize. Scientist and mathematician Gösta Mittag-Leffler, a previous Swedish proponent of the Curies for the 1903 prize in physics and now a backer of hers for the 1911 chemistry award, had sent six separate telegrams to her coworkers requesting her presence at the ceremony, otherwise fearing she would be considered guilty.[30]

Marie Curie was many things, a peerless scientist, a relentless investigator, a caring mother, a devoted woman to the man she loved.

But she was not someone who could comprehend being barred from receiving what she had rightly earned through her dedicated efforts. Her carefully penned response encompassed much of her belief in her rights, and herself, as a person of honor and integrity. Her duel was not to be fought with sword or pistol in order to defend that honor. Rather, it was to be fought as she had always done when faced with injustice, with logic and an enduring belief in principle. Her letter to Arrhenius stated: "The approach you are recommending to me would appear to me to be a serious error on my part . . . I consider that there is no relation between my scientific work and the facts of private life . . . I cannot accept the assumption that the appreciation of the value of scientific work can be influenced by libel and slander concerning private life."[31]

Marie Curie had withstood a month of withering target practice at her expense after such a promising beginning at the Solvay Conference with the euphoric news of her second selection as a Nobel Prize winner. The illustrious Svante Arrhenius, self-proclaimed Zeus astride Mt. Nobel, had now decreed her unworthy to receive another award until she met his personal standards. She was physically exhausted and mentally fatigued. Nevertheless, accompanied by her sister Bronya and young daughter Irène, Marie defiantly made the multiday trip from Paris to Stockholm, arriving in time to attend the Nobel award ceremony on December 10 and the banquet that evening, before her acceptance speech the next day.

Among the many attendees at the banquet was the widow of the Swedish scientist Knut Ångström, a faithful supporter of the Curies on their way to becoming Nobel Prize laureates in 1903. She had written a short note the day before, delivered to Marie before the evening's festivities: "I look forward to seeing you again in my life. That's why I'm taking part in the Nobel party banquet tomorrow night. I will never forget that spring day you spent with us in Uppsala . . . My granddaughter Dagny especially asked me to pay you her compliments. See you tomorrow! Your friend, Hélène Ångström"[32]

Hélène had been there in 1903, when her husband Knut had supported the Curies for a Nobel Prize in physics for their discovery of radioactivity and Pierre had to lobby for Marie to be given a portion of it. They had met in the university town of Uppsala in the spring of 1905, when the Curies, who had missed the actual awards ceremony in December 1903 due to illness and fatigue, traveled to Sweden to belatedly receive their prize. At that time, Hélène and Marie had gotten to know and admire one another a bit as individuals. Now, years had gone by, and Marie was still fighting to defend her rights, this time to another prize. Scientific society had attempted to deny her both prizes because she was a woman: her first award, simply because it was impossible to consider that a woman could be the equal of a man in the scientific discovery process, and the second, because it was equally unheard-of for some to accept that the professional efforts of a woman should be considered separately from the private life of that woman. Both ideas proved to be wrongheaded in their approach. Hélène Ångström knew it, and she felt it was important for others, including her granddaughter, to understand it as well.

The day of the awards found Marie still feeling weak, but her voice gathered strength as she gave her Nobel Prize acceptance speech in front of the king of Sweden, assembled dignitaries, and guests. Her preamble to her review of one of the greatest discoveries in scientific history proudly acknowledged that she and her husband had been a wonderful team that had investigated radioactivity and discovered the elements radium and polonium. However, she took great pains to explain the singular role she had played in much of this work, especially the insights that she alone had brought to the understanding that radioactivity was an elemental phenomenon, distinct from anything else ever attributed to materials found in the world. As she told the captivated audience the story of her efforts of investigation and discovery, the question of how she could be undeserving of a Nobel Prize for her work, at this time or any other, disappeared as quietly as the wisps of an instantly forgotten dream upon waking.

Curie's extensive remarks began with some background, driving home the essence of the discovery and clearly noting her own specific efforts: "Some 15 years ago the radiation of uranium was discovered by Henri Becquerel, and two years later the study of this phenomenon was extended to other substances, first by me, and then by Pierre Curie and myself. This study rapidly led us to the discovery of new elements, the radiation of which, while being analogous with that of uranium, was far more intense. All the elements emitting such radiation I have termed *radioactive*, and the new property of matter revealed in this emission has thus received the name *radioactivity* . . ."

"The history of the discovery and the isolation of [radium] has furnished proof of my hypothesis *that radioactivity is an atomic property of matter* . . . I believe that it is because of these considerations that the Swedish Academy of Sciences has done me the very great honour of awarding me this year's Nobel Prize in chemistry."[33]

Marie was precise in her description of why she alone was worthy of this Nobel Prize. Then, she graciously noted the contributions of her husband, caringly stating that a portion of the glory should rightly rest with Pierre, her partner in much of this scientific journey.

"I should like to recall that the discoveries of radium and polonium were made by Pierre Curie in collaboration with me . . . The chemical work aimed at isolating radium in the state of the pure salt, and at characterizing it as a new element, was carried out specifically by me, but it is intimately connected with our common work . . . the award of this high distinction to me is motivated by this common work and thus pays homage to the memory of Pierre Curie."[34]

Arrhenius should have understood. He had almost been denied his goal of entering the scientific community those many years ago when his doctoral dissertation was met with silent derision by those sitting in judgment. Yet, he had continued to pursue his quest, searching for the principles and meaning of the mysteries of the physical world and being rewarded with a Nobel Prize for his continued efforts based on refinements of this once-contested thesis. Years later, unintentionally perhaps, Arrhenius had now become what he had detested so long ago. He was presiding with his own arbitrary, biased judgment, in this case conflating professional scientific accomplishment with private, personal choices. All that was really appropriate was applying objectivity in rewarding valiant effort to discover secrets of the universe in all their complexity, no matter who had pursued these scientific truths.

Less than three weeks after her triumph at the awards, on December 29, 1911, Marie Curie was rushed to the hospital, her body and mind completely surrendering to the intense strains of the past sixty days. Her physical condition was dire, a serious kidney infection requiring immediate attention and an eventual operation. Her mental and emotional state were in scarcely better shape. Her relationship with Paul Langevin was to completely collapse as a result of the very hostile, public exposure of their affair, and the barrage of insults directed at her in the press had taken its toll. The winning of a second Nobel Prize was small consolation for her dreadful lack of well-being. She had paid a tremendous cost for being a woman of principle, her health deteriorated and dignity shredded. It was a price that would never be fully recovered.

CHAPTER TEN

A Different Dimension

The only father-son pair to ever jointly win a Nobel Prize received the award for physics in 1915. Two transplanted Australians living in England, University of Leeds physics professor William Henry Bragg and his son William Lawrence Bragg, a student at Cambridge, had found that the energy waves of X-rays, by their diffraction when shot through a crystal, resulted in a pattern captured on a photographic plate that was unique to the crystal's atomic structure. X-ray crystallography, as it became known, could be utilized to identify the three-dimensional composition of elements of matter. This understanding could, in turn, explain physical phenomena concerning the materials being examined. As an example, diamonds and graphite, both composed of carbon atoms, demonstrated two very different X-ray diffraction patterns. Diamonds were shown to exhibit a sequence of atoms with a very sturdy structure, resulting in great hardness of the material, while graphite's relative softness was explained by its atoms showing a layered arrangement, allowing each layer to be easily separated from the others. Eventually, this use of X-rays in examining chemical elements and compounds was to revolutionize the manner in which materials were characterized, to the point that when, years later, the unique double helix structure of the DNA molecule was identified, it owed its discovery to the powerful analytical tool of X-ray crystallography.

Laying the groundwork for the Braggs' discovery, the German physicist Max von Laue had won a Nobel physics award for his work in demonstrating that X-rays could be diffracted by crystals. In fact, during the years 1914–1917, X-rays utilized in various optics investigations accounted for all of the Nobel physics awards, in the process helping recapture a high profile position for X-rays in the public eye. Their discovery almost twenty years before by Wilhelm Röntgen, for which he received the first Nobel Prize in physics in 1901, was the advent of a popular craze about these mysterious energy waves and the skeletal "shadowgrams" they could generate of the human body. But, after a few years of fascination with the concept and the resulting images that it seems most everyone wanted photographers to take involving X-rays of their hand, if not other portions of the body, the use of X-rays had settled into the province of medicine as another professional means of examination and diagnosis for the physician. Just as important in the early 1900s was the developing use of focused X-ray beams as a radiation oncology treatment for some forms of cancer, soon to be augmented by the Curies' newly discovered radioactivity, as both were found to have selectively destructive effects on many types of diseased matter within the body.

During the summer of 1912, while the Braggs were vacationing on the Yorkshire coast in Northern England, contemplating the construction of experiments with X-rays that would uncover new secrets about the world around us, Marie Curie came to England's shores as well. She had managed to steal away from the depths of her despair in France to the coast southwest of London for a retreat offered by a female scientific peer, Hertha Ayrton. They had first met years before, in 1903, when Pierre Curie was invited to review the subject of radioactivity at a meeting of the Royal Society of London and was politely asked to bring Marie along as a guest. Their friendship grew with the years, Ayrton often defending Marie's part in the discovery of radioactivity. She famously responded with a pointed pronouncement to one subsequent article written in the English press that suggested Marie had ridden the coattails of Pierre in their work together, "Errors are notoriously hard to kill, but an error that ascribes to a man what was actually the work of a woman has more lives than a cat."[1]

Marie was depressed from the disastrous ending of her affair with Paul Langevin and the public attacks that had cast a cloud over her Nobel Prize win, as well as being physically debilitated from her recent kidney operation. She took her two daughters and, thanks to Ayrton's clandestine arrangements, spent a few secluded, restful summer months hidden from the public along the Hampshire coast of England. She had always enjoyed the outdoors, especially the ocean, and was thankful for this escape from what had become ever-increasing, unwanted notoriety that fueled the destruction of her reputation in France. More than ever, Marie craved the serenity of an uneventful few months with her girls and good friend. Watching the hypnotic motion of the waves lapping the shore, seagulls squawking in the background, was a tonic to revive her spirit. As her daughter Eve remembered it, "[Madame Curie's] friend, Mrs. Ayrton, received her and her daughters in a peaceful house on the English coast. There she found care and protection."[2] Even with this desperately needed respite, it was to take the remainder of the year for Marie Curie to regain enough strength to resume her work on radioactivity. During 1913, her plans to develop a Radium Institute in Paris, devoted to radioactive research and experimentation, were well underway. Then, the serenity of a Europe at relative peace for over a generation was shattered, setting the world wobbling on its axis.

The First World War, later naively termed the "War to End All Wars," broke out in August 1914. Sweden quickly claimed neutrality for the duration of the conflict, which might have allowed the Nobel awards to proceed. The prize in physics was planned to be given to the German scientist von Laue in Stockholm in 1915, by which time everyone presumed the war would be over. But the fighting did not cooperate with this schedule and all award presentations were subsequently delayed until after the hostilities had ended.

The widening conflict was to draw most of the rest of Europe, and eventually much of the world, into its path of devastation. As Barbara Tuchman described it in her Pulitzer Prize–winning book on the subject, "Europe was a heap of swords piled as delicately as jackstraws; one could not be pulled out without moving the others."[3] During the previous decades, an intricate web of treaties and agreements had been signed between various individual European nations to provide defensive support for the other if their countries were attacked by an aggressor. The Triple Alliance between Germany, Italy, and Austria-Hungary had its roots in military affiliations dating back to the 1880s. It was balanced with the more recent turn-of-the-century set of agreements that formed the Triple Entente of France, Russia, and England. As well, numerous smaller states had received assurances of military backing by some of these larger countries. The very existence of all of these alliances was meant to prevent major confrontations, but instead they acted to ensure any misstep would set into motion the wheels of war. Once ignited by the assassination of Austria-Hungary's Archduke Franz Ferdinand from a nineteen-year-old Serbian nationalist's bullet in late June, the flames of conflict quickly escalated into a howling firestorm due to the military responses dictated by these compacts. What developed was the first major war between nations of an industrialized world, with staggering human consequences, accounting for close to twenty million deaths and as many or more wounded.

Offense, lightning quick, relentless in force, and decisive in battle were watchwords engrained in the German military, following precise plans laid out for just such an eventuality years before.[4] Faced with the potential of twin fronts of war, on their eastern border with Russia as well as with France to the west, Germany had designed a strategy to attack first to neutralize the French in a matter of six weeks before turning back to grapple with the lumbering Russian bear. They almost succeeded, sending a massive force that butchered a neutral Belgium that lay in its westward path. Ideally, they would unite with the remains of a smaller southern wing that was meant to tempt the French into focusing their major offensive in the Alsace-Lorraine area, toward the center of the Franco-German border. Following this blueprint, the Germans were determined to triumphantly take their first stroll along the Champs-Élysées in Paris before autumn. But with timely aid from the British, the French were able to halt the German offensive just short of their goal during the first Battle of the Marne at the end of the summer. The fighting, which all participants had confidently proclaimed to their citizens would be over in a few months, now settled into years of trench warfare. Monstrously high casualties ensued for both sides, a stalemate where each was now determined to overwhelm the other, if not in decisive battle, then with destruction by a million cuts.

Not only had the Allied forces and Central Powers now been handcuffed into a protracted death match; new industrial means brought the conduct of large-scale warfare to another dimension. Often over the greater part of the past millennium in Europe, a military force representing an adversarial city-state, then country, laid siege at the gates of a foe's castle or walled city. But periodically, as time passed and larger armies marched far from their own homes, opponents crossed paths and lined up on either side of an open field to address their grievances in great battle. In coordinated waves of attack, for the most part they sent one force directly into the face of the other, masses of infantry wildly charging, often accompanied by mounted horsemen galloping at breakneck speed into the enemy's lines. The result was a chaotic scene of desperate, ferocious hand-to-hand fighting with sword, knife, pike, or axe. Later, this was replaced by gunfire, opening rounds of field cannon being discharged in an attempt to inflict initial casualties and weaken the other combatants before the inevitably ordered charge of soldiers with bayonets fixed on their muskets for closer fighting. But even as the weapons of war had evolved, what appeared to be relatively constant was the "irresistible force meets immovable object" scenario most of these battles created, thousands of soldiers from each side struggling mightily in individual combat with an equally determined enemy until one army's leaders blinked, gave way, and disengaged to flee the field, leaving the other victorious. The general conduct of this macabre game had gone on relatively unchanged in Europe for centuries, at least until the Industrial Revolution produced much more destructive means for combatants to maim and murder each other.

In the stalemate that resulted from the repulsion of the German juggernaut at the Battle of the Marne, both sides of the conflict now settled into trenches they dug for protection as well as to mark their territory. Long, thin lines of freshly dug earth, interspersed with makeshift underground shelters, connected the troops that lived there, waiting for orders to advance toward the enemy, who similarly were burrowed into their own deep etchings in the ground not far away. The glorified ditches were now employed to provide an earthen defense from bombardment by huge artillery pieces that were stationed miles behind the battlefront. These gargantuan machines could lob tremendous bursting shells into the enemy lines, leaving great craters where they landed and spraying chunks of hot shrapnel from the explosion to embed into whatever was in their targeted path, especially soldiers in the immediate vicinity. Airplanes, an invention of the new century, were beginning to be used not only in reconnaissance missions over hostile territory but as offensive weapons to drop exploding bombs on the soldiers below as well. And when the time came for one army to emerge from their labyrinth of earthworks and mount a spirited charge toward their foe, deadly machine guns employed from the comparative safety of the defenders' trench network were usually too much for the advancing troops to overcome. Bullets flew everywhere at once, usually sufficient to repulse the invaders and send them scurrying back to their own lines, often leaving piles of dead or severely wounded in their wake.

Early in the fighting, the French army was particularly susceptible in this last regard. Their most recent experience in full-fledged battle was the Franco-Prussian conflict of 1870. At that time, they proved no match for their Prussian counterparts, who quickly occupied Paris and ultimately forced the French to degradingly cede their border provinces of Alsace and Lorraine to Prussia as a territorial price for losing the war. Since that time, France had held themselves in embarrassed self-contempt for their wartime performance, judging their own passivity as being directly responsible for their defeat. Their generals swore this would never happen again, deciding to embrace a more offensive posture for future conflict, embodied in an "attack rather than defend" approach. Their infantry would mount the charge, heroically leading with their sturdy bayonets, ready to gut the enemy at close quarters. Unfortunately, by 1914, this strategy of "pitting *élan* and cold steel against machine guns and rapid-fire artillery" was catastrophic, the French suffering close to three million casualties in the opening phase of the war, during 1914–1915 over two million wounded or missing in action.[5] The sheer numbers were beyond belief, as was the suffering of those who were wounded, who often harbored bullets from machine gun fire or bits of shrapnel stuck in their mangled bodies. This was an issue that needed immediate attention where feasible, at least beyond quick amputation of shredded appendages.

In the early days of the war, as the casualties began to mount, Marie Curie, a patriot to her adopted country despite how she had been treated, pondered how she might help France address this dreadful situation. She began to piece together the beginnings of a plan to use her scientific knowledge to aid her country. As the other anxious inhabitants awaited the outcome of the great battle to protect the city, Marie Curie was already turning her thoughts into actions.

She had long been intimately acquainted with the use of X-rays as a means to examine and diagnose patients. In Germany, a month after Röntgen's discovery a Frankfurt newspaper was already touting the immediate medical advances to come, explaining the ease with which doctors could now determine the state of complex bone fractures and pinpoint bullet and shrapnel fragments without painful manual manipulation and probing.[6] As Curie stated, "[X-rays] make possible the discovery and the exact location of projectiles which have entered the body, and this is a great help in their extraction. These rays also reveal lesions of bones and of the internal organs and permit one to follow the progress of recovery from internal injuries."[7] However, the X-ray machines themselves were large, heavy, and cumbersome, and they required readily available electricity to function. As such, they were most typically found in large hospitals located in major cities, far from the battlefields.

Now, during the first few months of the conflict, Marie Curie taught a number of volunteers how to operate radiology units that were vital to assisting the wounded during the Battle of the Marne, albeit at stations quickly built for use many miles behind the battle lines. This work would evolve over time into the "organization and implementation of radiologic and radiotherapeutic services" wherever they might be most needed by the French military.[8] As the war continued on, Curie was to instruct approximately 150 women in courses she taught in a Parisian hospital on how to become a radiological technician, much needed to support the machines that were strategically placed closer to the fighting but behind the front lines, in almost two hundred field hospitals spanning the Western Front.[9] Now, X-rays not only currently supported the Nobel Prize–winning efforts of a number of physicists from across Europe, but more importantly they had become the centerpiece in a plan devised by one of the world's most brilliant minds to save as many people as possible from the horrific consequences of mechanized warfare.

Until the beginnings of the 20th century, if those wounded in combat were not left to die often excruciating deaths on the battlefield as a result of their injuries, they might be hauled off the scene for care, taken well behind the immediate fighting lines. The weaponry of war had changed major battle injuries from slicing and stabbing caused by sword, knife, or pike to wounds from bayonets and musket fire. The effect of the musket ball, invariably bringing bits of clothing into the body along with the bullet itself, was a quick source of infection that most surgeons were unprepared to adequately address other than by amputation.[10] Thus, the military surgeons throughout the 19th century who did their work close to the fighting were more akin to branch-trimming arborists, hacking off readily identified damaged limbs rather than examining the wounded more closely for intricate diagnosis and delicate repair. Then, as trench warfare progressed in 1914, casualties from flying bits of metal, either bullets or shrapnel, became commonplace. Armed with more advanced medical knowledge of bacterial infection of these wounds, most of the amputations that might have routinely taken place could be avoided if, within six to twelve hours after the time of the wound, the location and removal took place of "all missiles and especially the dirty and infected clothing . . . in finding [metal shards] which the X-rays are of the utmost service."[11] With field hospitals stationed well behind the front, the closer X-rays could be brought to the battlefield, the higher the probability that this medical tool could be used in a timely manner to prevent unneeded amputation and potential unnecessary death from infection and related trauma from amputation.

However, as in most every skilled profession known to man, those who claimed themselves to be the experts were often slow to adopt new ideas, clinging to "accepted practice" over innovative thought. Thus, many battlefield surgeons were often hesitant to readily accept radiology as the valuable military tool it was to shortly become. As a group, they were widely untrained in the emerging technology, leading to a higher comfort level in amputation or, at best, exploratory surgery for eventually finding metal fragments in the body rather than employing more accurate pinpointing offered by the X-rays. Yet, an article at the turn of the century in the English journal *Lancet* claimed, "if [X-rays] could be taken on the battlefield, they would give surgeons an incentive to operate under adverse conditions."[12]

Beyond instructing others in the establishment of mobile X-ray operations behind the lines of engagement, Curie was determined to bring these valuable diagnostic tools as close to the point of most need as she could. In order to do this, she distilled her requirements down to a few very basic items. First and foremost was an X-ray machine, photographic plates, and ancillary equipment that were all compact enough to be transportable in a vehicle. Next was the vehicle itself, appropriately outfitted with means to generate electricity as well as darkroom photographic developing capabilities. Finally, adequately trained personnel were needed to operate the X-ray equipment, accompanied by a driver to bring all of this to the places that most needed it. As Marie thought about each of these components, she knew that she had the knowledge to solve most of issues, and felt she could raise enough money to address the rest.

Scouring the laboratories of Paris, Marie was able to pull together the basic equipment needed for a smaller, portable X-ray unit that could be moved by automobile, convincing the French Red Cross to fund the purchase of the materials along with a vehicle that could be configured appropriately.[13] The electricity required to produce the X-rays would be generated from the vehicle itself, the petroleum-fueled engine sending power to a small generator to produce the current required. Trained personnel to run the X-ray unit, along with the driver, were essential to the mission. Initially, Curie herself had learned how to operate the equipment, but she needed to employ a chauffeur to take her into the field, simply because she had never learned to drive. Eventually, she determinedly mastered this task as well, even becoming adept at basic car repairs that might be needed if a sudden automotive breakdown occurred while driving to the front. Her daughter Eve described her mother unabashedly taking on the duties of a seasoned auto mechanic no matter what the circumstance, "energetically turning the crank of the recalcitrant motor . . . putting her weight on the jack to change a tire, or cleaning a dirty carburetor with scientific gestures . . ."[14] During the course of the war, Curie was able to solicit enough funding to put twenty such vehicles into operation, bringing X-rays where they could be of maximum service to the military.

Thus was born the "petite Curie," the name given by the French public to honor the compact mobile X-ray unit and vehicle Marie Curie devised and brought to the front lines to save countless wounded French and Allied soldiers. It had taken a world war for this ingenious, determined woman to reclaim her dignity from a nation that, only a few years before, had mocked her as an unscrupulous foreigner not fit to bear the name Curie as the wife of Pierre. Without a second thought to the undeserved treatment she had stoically endured, Marie dedicated herself to accomplish her goal with no other purpose than to apply her scientific background, and her indominable spirit, in service of a country that needed her now more than ever.

Marie's elder daughter, seventeen-year-old Irène, was just as relentless in her desire to accompany her mother as an able assistant in her work. She, like Marie, was extremely bright as well as single-minded in achieving her goals. In Marie's pursuit of her vision of providing patriotic wartime relief to wounded French soldiers she could find no more worthy a companion than her own daughter. After taking a nursing course in the first few months of war, Irène next learned about radiology from carefully observing her mother and the radiological technicians at work. Eventually, she became a radiological trainer herself in Paris when not running her own X-ray unit for a time during the conflict at a field hospital in war-torn Belgium. All the while, she continued her education at the Sorbonne when she could be back in Paris, receiving multiple scientific and mathematical degrees as the war dragged on.

As Marie Curie's idea to take an X-ray machine right to the very brink of the fighting gained credence with the French military, Germany was similarly exploring innovation along the same lines in support of its troops. Röntgenology, as opposed to radiology, was the preferred German term for use of X-rays, honoring its Nobel laureate discoverer, Bavarian Wilhelm Röntgen. The intriguing new technology had received little skepticism from the German medical community, which in the years following its invention rapidly embraced the new discovery as a valuable tool, especially for surgery.[15] In the early war years, vehicles similar to the "petite Curies" could be found in some of the front lines of the German forces, portable Röntgen-ray machines and equipment being utilized, with electricity for the machine being supplied by a generator hooked up to the engine of a motorcar or, in some cases, even a motorcycle.[16]

Generally, the German mindset toward scientific discovery and innovation could be considered second to none in Europe at the beginning of the 20th century. They were proud of their institutes of higher learning and the venerated "Herr Professor Doktors" who led their programs, especially in physics and chemistry. Röntgen, Planck, Wien, Nernst, and Ostwald were revered names in German physics and chemistry teaching halls across the country, all of them having won, or were soon to win, Nobel Prizes in their fields of expertise. Germany's scientific prowess and related industrial achievements could be attributed in large part to this focused educational system, churning out theorists, researchers, and engineers to staff the developing industries employing this knowledge base and skillset, including dyestuffs, pharmaceuticals, fine chemicals, electrical devices, and scientific instruments.[17] Ostwald, the 1908 Nobel laureate in chemistry, was widely quoted in a Swedish newspaper during the war as pinning German scientific success to a particular characteristic that he felt was almost unique to the Germanic race, organization.[18] Certainly, many might have agreed that the Germans possessed this capability, but few, including Marie Curie, felt Germans alone had cornered the market on it.

Not only was science held in high esteem in Germany; the scientists themselves for the most part had as patriotic a view of the superiority of their country as they did of their scientific accomplishments. To this end, in October 1914, ninety-three scientists and intellectuals signed a document that was to forever link their names with the atrocities of Germany in its early conduct of the war. Dubbed the "Manifesto of the Ninety-Three," the proclamation indignantly refuted the supposed untruths being spread around the world about the start of the war by Germany and its subsequent conduct. It advanced the notion that Germany should essentially be held free of blame from the actions taken during the opening phase of the war, especially the brutal violation of the neutrality of Belgium, claiming it would have been suicide not to have taken this preemptive move.[19] Each denial was prefaced with the statement "It is not true," followed by various accusations made by the Allied nations and accompanied by tortured logic as to the untruth of each action that Germany actually took. It ended with the plea "Have faith in us! Believe, that we shall carry on this war to the end as a civilized nation, to whom the legacy of a Goethe, a Beethoven, and a Kant, is just as sacred as its own hearths and homes."[20] This, from a country that was soon to introduce the chemically toxic vapors chlorine, phosgene, and "mustard gas" to the World War I battlefield.

Besides the near-mythical Röntgen, among the scientists affixing their names to the statement were German Solvay conference attendees Max Planck, Walther Nernst, and Wilhelm Wien. Planck was an old world traditionalist, brimming with hope for his country. His signing of this document, which he was soon to regret and at least partially recant, matched his rather outspoken feelings of Germany's rightful place at the head of European nations. Nernst, ever the innovative scientist, participated for a time in the development of chemical agents for gas warfare. Exceeded only by losing one's own life in battle, both suffered the ultimate sacrifice of the war they felt was being righteously waged by their Fatherland, Planck's eldest son and both of Nernst's boys joining the army and dying in the conflict.

At the time, a notable newcomer to the German scientific community, another participant at the Solvay Conference, specifically avoided signing the manifesto. This was Albert Einstein, newly appointed professor at the Prussian Academy of Sciences and the University of Berlin, who had arrived from Switzerland with his family in the spring of 1914. Planck and Nernst had been so favorably impressed by the young physicist at the conference that they had worked industriously almost from its conclusion to construct just the right position to lure him from his post in Zurich and secure him as a gleaming jewel in the crown of German scientific achievement. Their strategy was to devise a position that, though prestigious, required no actual teaching or administrative duties while paying him the top salary a professor could legally earn at the time, 12,000 deutschmarks.[21] This way, Einstein could devote his full efforts to the theoretical machinations that had already made a mark on the world of physics, and on whom Planck and Nernst were gambling would continue to build on these unique insights, to the greater glory of Germany. Einstein certainly enjoyed the attention but felt the pressure to perform that came with it. He spoke to a friend about this situation, comparing himself to a prized fowl and hoped he could produce an egg on demand.[22]

In being offered the opportunity to become part of the German scientific elite, it probably never occurred to those already enveloped in its patriotic stirrings that Einstein would not act as a proud German once war had begun, as they had when they signed the Manifesto of the Ninety-Three. However, although Einstein had become a German citizen in accepting his position at the Prussian Academy of Sciences, deep in his heart he remained an internationalist. He had renounced his German citizenship as a birthright many years before when leaving Germany rather than serve in the army, but when he returned to Berlin he was also allowed to retain the Swiss citizenship he had collected in the interim. He was a pacifist, though he touted these virtues to only his closest friends. Now those held in highest esteem in German science, including the two pillars Planck and Nernst, who had recruited Einstein to his new post, were signing a petition that he could not endorse.

Einstein's reaction to the document went further than not signing it. In October, along with fellow physicist Georg Friedrich Nicolai, Einstein produced a proclamation that came to be known as the "Manifesto to Europeans." It was his first outward statement to the world concerning his political leanings. The document suggested that the people of Europe look at themselves as part of a greater heritage and work together as one to support all its inhabitants. It predicted that the war would only produce an even more fractured world, humbly imploring that the terms of peace be reasonable in dictating the new European order so as not to plant the seeds of future conflict.[23] This request astutely presaged the codification of the Treaty of Versailles at the end of the war, with its crushing financial burden of reparations placed on Germany along with territorial concessions and the psychological stigma of world condemnation. When Einstein and Nicolai circulated their manifesto to those they felt might be amenable to signing it, only two others responded positively, causing them to abandon the larger publication of the document. Nicolai's career was effectively ended with the negative publicity in Berlin as a result of the proclamation. Einstein's continued relatively unscathed. The golden goose was still expected to deliver its precious eggs.

Einstein's world was moving in many different directions as the war began. His acceptance of the Prussian professorship was the latest in the continuing mercurial ascent his career had taken since the autumn of the Solvay Conference. In each successive move, from Prague to Zurich and finally settling in Berlin, Einstein's path had been marked by those who stood on the highest rungs of the professional physics ladder encouraging and enticing him to hurriedly make the climb to join them. Curie's and Poincaré's effusive recommendations had paved the way for his move to Switzerland, while Planck and Nernst had followed shortly thereafter by all but offering the post in Berlin on a shiny silver platter. Yet, at the same time, as his professional standing increased dramatically, his deepest thoughts on the structure of society were in conflict with the nationalism swirling around this new position. To add to the complexity, his marriage was disintegrating in a chilling manner.

Two people in love together can overcome many obstacles on life's journey. Frequently, the climb secures their bond even more tightly, two adventurers leaning on each other for strength and comfort. Just as often, the challenges can tear this delicate connection apart, inflicting irreparable wounds that accumulate over time until the relationship breaks. Sometimes, either one or both parties find the spark is no longer there. In Einstein's case, he now appeared to view his marriage to Mileva Marić as shackling him to an unwanted future. The move to Berlin presented an opportunity to cleave this burdensome chain once and for all.

The couple had once been madly in love when they met at a Swiss technical university, sharing their intellectual dreams and earthly desires. A baby girl had come from their pairing, an innocent child whose fate was sealed in her illegitimate birth and possible scarlet fever–related death as an infant—or covert adoption by another. Once married, they had put thoughts of her behind and raised two boys that were the center of their world for a time. Einstein's jobs took them to Bern, Zurich, Prague, and Berlin, Mileva preferring Switzerland, where she had felt comfortable as a student as well as a young mother and wife, having made friendships there. Now, coming to Germany with her sons to be with her husband, Marić's unease quickly turned to distress with Einstein's seeming disinterest in her arrival. She learned that her husband was involved with another woman, his cousin no less, and that Einstein's only concerns appeared to be his new job, his new love interest, and making life uncomfortable for Mileva while trying not to endanger his relationship with the boys.

The finale of this sad one-act drama in Berlin came in a formal memorandum, delivered from Einstein to Mileva by mutual friends in mid-July 1914. Its contents were clearly stated in a crisp, sterile manner. Examining it has the feel of discovering an awkward contractual agreement between a master and a servant/companion, which seemed to be Einstein's intention. The document laid down the completely one-sided conditions under which he would accept remaining a partner with his wife.

A. You make sure

1. that my clothes and laundry are kept in good order and repair
2. that I receive my three meals regularly *in my room.*
3. That my bedroom and office are always kept neat, in particular, that the desk is available *to me alone.*

B. You renounce all personal relations with me as far as maintaining them is not absolutely required for social reasons. Specifically, you do without

1. my sitting at home with you
2. my going out or traveling together with you

C. In your relations with me you commit yourself explicitly to adhering to the following points:

1. You are neither to expect intimacy from me nor to reproach me in any way.
2. You must desist immediately from addressing me if I request it.
3. You must leave my bedroom or office immediately without protest if I so request.

D. You commit yourself not to disparage me either in word or in deed in front of my children.[24]

It took only eleven days after receiving this curt dismissal of their eleven years of marriage for Mileva to be on a train back to Zurich with the boys, realizing this was no business deal of which she wanted any part. The two had agreed to separate, working out an acceptable financial arrangement including division of property and monthly allowance without taking the final step of divorce, which would happen shortly after the war in 1919. Einstein, though momentarily overcome with emotion at their departure, quickly went back to his professorship and romancing his cousin, with whom he had been secretly corresponding for two years. He had an ability to compartmentalize his personal emotions from his other thoughts about the puzzles of the physical world, which at times seemed to overflow, continuously churning in his brain. Once he had achieved his goal of severing his daily relationship with his wife, albeit at the expense of losing constant contact with his sons, he could turn back to another pressing struggle in his mind, the general theory of relativity.

By August 1914, the war was ramping up on the Western Front and his fellow professors were becoming enthralled with the idea of a quick German victory march in Paris. However, Einstein's gaze was centered in the opposite direction, east of Berlin toward the Crimea in Russia, where he knew he would find the proof of his theoretical concepts. Specifically, he was anxious for news about a total eclipse of the sun that would happen during late August, which could best be observed in Crimea and could verify the crux of his theory. The details of Einstein's thinking on relativity had all been coming together not only after years of intense effort but more recently with mathematical equations he had developed with the help of his good friend and mathematician Marcel Grossman that led to very definite conclusions about gravity and light. He felt confident in his work, sensing he was about to deliver to Germany the first, and perhaps proportionately the largest, of many golden eggs.

Einstein had been grinding away for some time now, steadily constructing an almost incomprehensible idea to others that he had harbored for quite a while about space and time, gravity and light.

Building on the principles he had expounded almost a decade ago in his "miracle year," he had come to view space as not separate from time but interwoven with it into something he and a very few others now termed space-time, a fourth dimension that could be affected by large objects within it, such as the sun. Traditional physics was thought of in the three-dimensional terms of space, objects moving with inertial velocity, front to back, side to side, and top to bottom. Within that context, Newton had developed the overriding principle of gravity that acted between all bodies of matter, especially observable in the relationship to large bodies like the sun and its orbiting planets. Einstein now introduced another dimension into the picture, an interrelationship of space and time. Conceptually, if space-time was thought of as a sheet held taut and stretched across the infinite expanse of the universe, it was perfectly flat in all directions. But if a large object like the sun existed in space-time, it created a depression or curvature in the sheet. The larger the object, the greater the curvature. Einstein's view of gravity was the depression created by the large body of the sun existing in the space-time dimension, the curvature caused by the sun acting as the "gravity" that compelled the planets to orbit around it. He reasoned that light near the large object inevitably followed this curvature as well that was created in the dimension of space-time. By contrast, Newton, and everyone else for over 250 years, thought only in three dimensions as they contemplated gravity. As such, they agreed that a large object like the sun exhibited a strong gravity inherent to large objects. Newton, who had hypothesized that light was made up of tiny particles he called corpuscles, felt that gravity would be strong enough even to bend the light closer to it by just a tiny fraction, but nothing like the effect of the sun on light when viewed in the context of the space-time dimension.

Einstein's challenge lay in demonstrating in the mindset of a three-dimensional world that a fourth dimension also existed. He could think of only a few ways to do this, the most promising of which incorporated the natural phenomenon of a total solar eclipse. Einstein reasoned that his theory could be given finite validation during the total eclipse if photographs could be taken from precisely the right location at just the right moment so that starlight could be shown to be bent to the degree he had calculated as it passed the sun. Pictures taken of the sun would be the key, with starlight right on the periphery of the sun's outline as the total eclipse occurred, the rays bending from the normal position of the stars and showing the expected deflection of light that his theory predicted if he was correct.

Total solar eclipses were relatively uncommon events and as such were usually sought out and observed by astronomers whenever possible. The most recent eclipse prior to the anticipated one in 1914 had been in 1912. At that time, a British expedition headed by Arthur Eddington had traveled to Brazil to make its observations, making the long journey only to be denied the reward of viewing the spectacle by the bane of astronomers, thick cloud cover and heavy rain.[25] Now, the best place on the planet to observe the total solar eclipse of 1914 would be on the Crimean peninsula, and Einstein had worked with an astronomer acquaintance in Berlin, Erwin Freundlich, to plan an expedition there for this very purpose.

Unfortunately, as the German expedition arrived in early August and began to set up camp, positioning their equipment for its observational and photographic purposes, the war was gaining importance in Russia's activities. Suspected of being spies sent to monitor Russian movements at their large naval base in the area by employing their specialized optical equipment, Freundlich and his compatriots were taken into custody by the Russian authorities.[26] The expedition had ended before its intended mission had even begun, leaving Einstein's definitive proof of his general theory of relativity to wait until the next full solar eclipse after the war, due to take placc in 1919. Arthur

Eddington was to again lead the observation team, this time successfully photographing the event and with it providing confirmation of Einstein's theory to an astonished world.

The summer of 1914 had been one of great tragedy for Europe. The underpinnings of society had come undone in spectacularly horrific fashion, resulting in a war that would badly scar its participants and leave its survivors to contemplate if there might be a better way for civilization to exist going forward. The Treaty of Versailles would be hammered out by the warring parties, gaining self-serving satisfaction for the victors but having a pernicious effect with the losers that would culminate in another world war less than a generation later. After the treaty was signed, world leaders began to discuss a global organization that they envisioned would work to prevent war on this scale from ever happening again, forming the League of Nations with this fervent hope. Notably, the United States, whose late entry into the conflict had turned the tide against Germany, never became a member nation of the League. Without America's participation, it turned out to be a well-intentioned but relatively powerless body in the prevention of world conflict.

Marie Curie and Albert Einstein had approached that summer from two different perspectives, one working within the traditional three dimensions, the other lost in complex thought and calculations that predicted a fourth. Curie had spent her time in service to her country, conceiving and executing a creative plan to employ science in a manner that could repair a multitude of soldiers' bullet- and shrapnel-ridden bodies by the timely use of compact X-ray technology. As always, she had approached the challenge with logic and unstinting effort, the combination again proving successful for her as it usually had in the past. Einstein had seen war from another viewpoint, when he allowed it to permeate his thinking at all. He could scarcely believe his German scientific colleagues could so wholly embrace the militaristic fervor wrapped in unvarnished nationalism that had permeated Germany's very existence. But the coming war, as well as the collapse of his marriage, appeared to be little more than irritating diversions from his relentless quest to confirm an understanding of the natural world that would forever change the principles on which it was based.

CHAPTER ELEVEN

After the War

If someone lived in the United States during the spring of 1921, it was almost impossible not to hear of, or perhaps even witness, the scientific phenomenon that was happening across much of the country at that time. The daily newspaper coverage of events was filled with enthusiastic descriptions of the situation. Within a month of each other, the two towering figures in the world of science, Albert Einstein, then Marie Curie, separately made their first ever trips to America. Einstein arrived in early April and was gone by the end of May, while Curie landed in mid-May and left for home at the end of June. Both visits were accompanied by tremendous fanfare wherever these two luminaries traveled, crowds showering them with adoration the likes of which was rarely seen. It's surprising that in their overlapping time in the country they didn't have the opportunity to meet at some common juncture, but apparently their schedules were so tightly choreographed that it wasn't possible. If they had, there's no telling how supercharged the atmosphere would have become when Radioactivity met Relativity.

Curie was a household name by then based on her discovery, with her husband, of radioactivity and the cancer-treating element radium over twenty years before, subsequently winning two Nobel Prizes for her, one in physics and one in chemistry. The intellectual insight that hypothesized atomic decay as the source of radioactivity was so novel a concept that the virtually exclusive male domain of science could scarcely believe it had emanated from the thoughts of a woman. Einstein, eleven years Curie's junior at forty-two, was already a deity among the scientific community but had only recently achieved worldwide notoriety a few years before with the much-publicized proof of his general theory of relativity during a solar eclipse, when his unconventional hypotheses about gravitation were proven with photographic evidence of the bending of light by the sun. Six months after his American journey he would finally be awarded his own Nobel Prize in physics for his monumental theoretical scientific achievements.

Einstein had primarily come to America to solicit donations in support of an as yet unbuilt Hebrew University in Jerusalem, while Curie had traveled to thank the women of America for raising funds to buy the precious radium she required to keep her radiological research effort afloat. They hopscotched across the country in a flurry of activity, both squeezing in time to meet the newly elected president Warren Harding in Washington, D.C. Curie was feted by half a dozen women's colleges and received honorary degrees at numerous other universities, sprinkling in receipt of honors from women's scientific associations with visits to scientific laboratories and a tour of the country's dominant radium manufacturer along the way. Einstein spent larger chunks of time in major Northeastern metropolitan areas and Chicago, mixing fundraising events with multiple lectures at elite scientific universities.

Everywhere they went, both were treated with a mixture of fascination and curiosity normally accorded visiting royalty, which indeed they were if one considered their stature as among the world's leading physicists. Einstein was the physics equivalent of Augustus Caesar, the conquering hero of relativity and newly crowned ruler of science, greeted with trumpets blaring and crowds roaring. Marie Curie was the scientific personification of England's first Queen Elizabeth, revered for the dedication to her craft that yielded the discovery of radioactivity and with it proof that women were indeed equal to men, at least in the professional realm.

Einstein was accompanied by his wife as their ship made port in New York in early April. They traveled in the company of Dr. Chaim Weizmann, a Jewish biochemistry professor from England who had emigrated from Russia and now headed the World Zionist Organization, dedicated to securing a permanent homeland for the widely dispersed Jewish community. Weizmann wanted to employ Einstein's newfound global celebrity as a magnet to attract potential funding for a planned university in Palestine. Einstein, though not a religious Jew, was witnessing an alarming rise in the already strong feelings of anti-Semitic scapegoating growing in the aftermath of Germany's defeat in World War I. As this awareness increased, so too did Einstein's desire to help his brethren in some significant capacity. This was that opportunity.

Since the mid-1880s, every ship that approached New York Harbor encountered a unique view like no other on earth. Soaring from the water off the tip of Manhattan was the Statue of Liberty. One hundred years ago, many encountering her for the first time had thought of her as the big woman with spikes on her head and torch raised proudly in her hand as a shining beacon to all. Over the years, the sculpture's outer appearance had transformed from its original reddish-brown copper color to a light bluish-green. Its finish had oxidized in the air and water to form a protective patina that worked to preserve her and the allegorical intention she was created to project.

The sculpture had come to beautifully symbolize to those arriving in this land that, no matter their past status, they were welcome in a country where they could find the freedom and opportunity to strive to attain their dreams. From the late 1800s through the early 1900s, millions had been enticed by the statue's message, fleeing the old world with unbounded hopes for the new. As they stared in wide-eyed wonder at Emma Lazarus's Mother of Exiles, it urged them to break the chains of the past, offering unconditional acceptance that only a mother's love could provide in the hope of nurturing a better future for all her children who sought it. On their maiden voyages across the Atlantic, Marie Curie and Albert Einstein each could surely attest to the Statue of Liberty's welcoming aura, but neither could have anticipated it as a harbinger of the generosity of the American spirit from which they would so greatly benefit.

As Einstein stood on the deck of his ship entering port, a stiff ocean breeze wreaking havoc on his already unkempt head of hair, he could scarcely avoid the dominant presence of the statue. Passing under its watchful gaze, he was following the path taken by a multitude of European Jews in the past decades, most fleeing anti-Semitism for the safety of America, a fate that would befall him as well a dozen years later. Many of them had seized the opportunities available in their new homeland, where ingenuity and hard work were a winning combination to unlock the pathway to a better life for themselves and their progeny. Einstein had come to request that some of them share a portion of their good fortune to secure a sustainable future for their fellow Jews in Palestine.

Curie, whose husband had died fifteen years earlier, landed in New York in mid-May with her two daughters, twenty-four-year-old Irène and sixteen-year-old Eve. Her trip was sponsored by the efforts of the American Marie Mattingly Meloney, known as Missy, the editor of the widely read women's magazine *The Delineator*. Missy had made the ocean voyage with the family, a companion for Curie to help steady her nerves in her first transatlantic crossing. Like so many passengers making the trip before her, Curie was made mildly ill from the rough crossing. Missy was no doubt sympathetic to her discomfort, in the process assuring her that what awaited at the end of the line was worth the temporary ailment.

Meloney had met and interviewed Curie the year before in Paris. During the course of their conversation she learned that the Curies, upon discovering the radioactive element radium and devising methods for producing it, had decided that for the good of mankind they would forgo any profits that could be generated by patenting these processes for manufacture, making production available to the world on an unrestricted basis, cost-free. Although only a tiny amount of radium was needed to support continuing experimentation with the substance, unbeknown to the Curies the cost of producing this material was becoming prohibitive. A gram of the material was exorbitantly expensive, worth over $100,000 in 1920. Her existing radium supply was now running out and Curie could not afford the cost of a gram of radium for continuing her experimental work. Missy decided to do something about it. She combined her promotional influence through her magazine with Marie Curie's worldwide reputation of ardent devotion to her work to successfully solicit more than $100,000 from women across America, enough money to buy a gram of radium for Marie Curie's continuing research efforts.[1] Missy wanted to present this gift from the women of America to Marie Curie in person, in the United States. Hence Curie's maiden sojourn in America was scheduled.

Both esteemed scientists made separate visits to meet President Harding, who had just been inaugurated in March of that year. Einstein came to the White House in late April as part of a contingent from a National Academy of Sciences gathering in Washington, D.C. Since neither spoke the other's language, the meeting was little more than a photo opportunity for the press—President meets Genius. When asked by journalists about relativity, Harding readily professed to know nothing about it, as did virtually all others who had been exposed to Einstein's concept. Even Dr. Weizmann had previously admitted the same ignorance, noting that on the voyage to America, Einstein's repeated attempts to explain the theory to him resulted in firmly convincing Weizmann that at least Einstein understood what he was talking about.[2]

Missy Meloney had arranged a mid-May ceremony at the White House to have Harding bestow upon Curie the gram of radium. The president presented her with an elaborate lead-lined box-and-key gift combination, the key to open the box where the radium would be stored. The radium was not actually there, but squirreled away for safekeeping until the Curies' voyage home the following month. Harding gallantly escorted Curie down the steps of the White House for photographs to commemorate the event, Curie flashing the hint of a *Mona Lisa* smile, her mind no doubt contentedly envisioning the securely stored radium gift she would soon have in her possession to continue her ongoing scientific investigations.

At the end of their stay in America, both Einstein and Curie were physically drained, having personally expended maximum effort to accomplish the goals of their journeys. So many people shook Curie's hand during the beginning weeks of her tour, one even injuring her in the process with a vigorous handclasp, that she finished her stay "with one wrist bandaged and the arm in a sling."[3] Whether in fact injured or bandaged just to avoid further glad-handing from the crowds, the frail, reclusive Curie was just not used to the crush of people she was expected to meet and greet.

Both scientists commented that they were satisfied with what their trips had achieved. Curie had received her coveted gift of radium bestowed from the generosity of American women while representing to the country the pinnacle of women's scientific achievement. Einstein had employed his recently found worldwide fame to modest but significant fundraising success, obtaining pledges for over $750,000 toward Weizmann's ambitious goal of multiple millions.[4] And, as scientific minds do, both had continuously observed the American people and culture along the way and had comments on what they saw.

Einstein expressed his initial impression of Americans' "joyous, positive attitude to life. The smile on the faces of the people in photographs is symbolical of one of the American's greatest assets. He is friendly, confident, optimistic, and—without envy . . . The American lives for ambition, the future, more than the European. Life for him is always becoming, never being."[5] Einstein sensed Americans' energetic acceleration into an opportunity-filled future versus Europeans' relative inertia. He felt there was a generally cooperative attitude across the population, affecting both business as well as academia, driving the country toward success in each venue. With all of this forward momentum, Einstein made an impassioned plea for the engine of progress that was America not to regress and divorce itself once again from the rest of the world.

America had only recently been dragged from its isolationist existence into World War I and had emerged as one of the most powerful countries in the world. In the years following the war, President Woodrow Wilson had literally given his last modicum of strength to help establish the United Nations' precursor, the League of Nations, in 1920. It was an attempt to engage the world's countries to band together to work toward peaceful harmony in the aftermath of the "War to End All Wars." Yet, even as the League's charter was being signed by a host of nations from around the globe, the United States chose to decline participation in it, an echo of George Washington's admonition to "beware of foreign entanglements" still ringing in the recesses of the country's psyche.

Einstein now spoke of the pressing need for America to stay involved in world affairs. "The United States is the most powerful technically advanced country in the world to-day. Its influence on the shaping of international relations is absolutely incalculable. . . . The last war has shown that there are no longer any barriers between the continents and that the destinies of all countries are closely interwoven . . . The part of passive spectator is unworthy of this country and is bound in the end to lead to disaster all round."[6] Einstein was a pacifist, as was Curie, their positions reinforced by having lived through the devastation that Europe experienced in the First World War. Both yearned for a world that actively sought to prevent warfare from ever flaring so violently again, and hoped the United States would take the lead in this effort rather than shy away from it.

Curie's observations of America and its people were somewhat similar in nature. Rather than publicly commenting, she penned her thoughts in notes and papers she sent to her friend and tour sponsor Missy Meloney. Her first observations were of the sheer physical vastness of the country. Due to Curie's mounting exhaustion as the tour wore on into its second month, Missy had canceled a week of travel and events on the West Coast and instead had Curie and her daughters take a respite and absorb the spectacular majesty of the Grand Canyon.

Curie was able to relax a bit on this diversion, taking in the profusion of colors and intricately etched terrain of her surroundings. She spent most of their stay leisurely in her hotel on the southern rim of the canyon, an enormous lodge within a dozen paces of the gorge's edge, offering a magnificent view. The young ladies were weary as well, not so much from the trip itself but from its formalities that had prevented them from really enjoying their stay up to this point. Now, temporarily, they could escape their bonds and explore the canyon on horseback as well as by taking the classic mule tour down to the river at the canyon floor.[7] One of the girls noted that there was nothing so spectacular with which to compare the canyon in Europe, while Curie's secretary said that it all emphasized to Marie the "spirit of the west and the great freedom of the out of doors."[8]

Before her western interlude, Curie had visited numerous women's institutions of higher education, from Smith College and Vassar to Bryn Mawr, Mt. Holyoke, and Wellesley. These uniformly left her in amazement at the differences between American and French education for women, even down to the logistics of student living. Curie noted the campuses in sprawling country settings, the cleanliness and comfort afforded the students in their housing, the comparatively large student bodies, many in the thousands, and the mixing of students from different backgrounds and social conditions. Perhaps most impressive to her was the general "joy of life" that was evident in the young girls, "the smiling and excited faces and in the rushing over the lawns to meet my arrival" which Curie would not forget.[9] All this contrasted with the relatively limited and repressed state of women's opportunities for education in France and Europe in general.

Bemoaned was the fact that not enough time in Curie's schedule could be devoted to visits to laboratories and scientific institutes, which were of great interest to her especially due to the long hours she herself had spent perfecting industrial production processes and utilization techniques for isolating the radioactive elements she had discovered. In this regard, she counted her visit to the Standard Chemical Company of Pittsburgh, Pennsylvania, at the time the largest processor of radium in the world, as one of the most important stops in her trip.

A train carrying Curie, her two daughters, and Missy Meloney arrived in Pittsburgh on May 25, met by various academic and political dignitaries, as well as many of the city's social elite, typical of her arrivals throughout Curie's tour. She had just received two honorary degrees in Philadelphia a few days before, a doctorate of laws from the University of Pennsylvania and a doctor of medicine from the

Women's Medical College.[10] Similar treatment followed on May 26 at the University of Pittsburgh, where she was presented with an honorary doctor of laws degree.

Pittsburgh and the surrounding area had a population strongly Eastern European in origin, including a large Polish immigrant community that wished to honor their fellow countrywoman. At a reception following the degree presentation, in a special welcome arranged by the Polish Alliance of Pittsburgh, an assembly of a group of little girls dressed in white held aloft the Polish flag at the entrance to the reception hall through which guests passed as they entered.[11] A souvenir of Pittsburgh was then presented to Marie by the contingent. As can be imagined, her heart must have been warmed by the gestures of her Polish brethren.

Before the reception, Curie had toured the laboratories of the Standard Chemical Company, which were located in downtown Pittsburgh. This company had produced the gram of radium that had been donated to Curie by President Harding in her White House visit earlier in the month. The next day, she was guided on the arm of Standard Chemical's president while taking the plant superintendent's tour of the company's manufacturing facility south of the city. She was in her trademark long black work dress and he wore a dapper double-breasted suit. Together they made a strange-looking pair for the industrial workplace. The air was pungent with traces of sharp chemical odors combined with white puffs of steam from processing equipment, the hard-packed ground underfoot a dusty mix of dirt and gravel. Yet, both were completely at home in this environment, content as if on a Sunday promenade in the park. They were kindred spirits, both proud of all of their chemical wizardry, from Curie's laboratory methods devised for the production of pure radium salts to Standard Chemical's scale-up of many of those same processes to a full-blown commercial enterprise. Curie's interest was piqued the more she saw of the facility. She enjoyed quizzing the tour-guiding plant manager on the many intricacies of his operations, feeling somewhat revitalized with these technical discussions, which were few and far between while in America.[12]

After her tour of America came to an end, Curie provided some written comments to Missy on a few of the other major cities she had seen in her travels:

"I was especially impressed in New York with the generous opportunities for free education, providing for both men and women . . . I have noticed that so much attention is given to public health, to playgrounds for children and the pleasures of the people. . . . [Washington, D.C.] is one of the most beautiful cities I have ever seen . . . I have especially admired the White House with its dignity and simplicity: a fitting home for the chief of a republic."[13]

Her thoughts centered on the opportunities afforded the population as a whole, down to the details concerning the physical well-being of its citizens. Her opinion of the unpretentious White House, juxtaposed with the opulent palaces of France and Europe, presented a picture of marked contrast for democratic government versus authoritarian rule.

Granted, Curie's favorable impressions of America were, in large part, centered on the slice of privileged life to which she was exposed on her visit. The bucolic campus settings of women's universities and their inhabitants as well as the almost reverential receptions she received across the spectrum of women's organizations who showered honors upon her were certainly a special portion of the American experience. Yet, in her fleeting observations of the urban life that were on display in New York City and Washington, D.C., she could sense the relative freedom and assistance given to those less fortunate, so different from the stiffly class-structured, old world order of her home continent. Upon embarking on her return voyage at the end of her visit, Curie commented, "I go back to France . . . with a feeling of affection for

[this] great country connected with ours by a reciprocal sympathy which gives confidence in a peaceful future for humanity."[14]

Marie Curie and Albert Einstein, two of the greatest scientific minds of all time, had each come to America on their first visit in 1921 on separate missions, at the same time having the opportunity to observe the newly emergent technological and financial giant that was America. Although they had been somewhat annoyed by the atmosphere of celebrity that accompanied their every appearance, each had incisively perceived something very special about the people in these crowds. Both were quick to note the opportunities that arose from the shared characteristics of optimism, hope, and freedom permeating the diverse population. They were touched by the inclusive nature of its inhabitants, their can-do attitude and vision of a brighter future with the opportunities available to them. They saw a land of ambition, where the collective whole overshadowed the individual, where there was a social conscience attached to the accumulation of great wealth. They hypothesized that all this formed the nexus of American greatness, for the benefit not only of the country but as an example to nations across the globe. A country that they felt, for the good of itself if not for the good of the world, needed to assume the mantle of global leadership in pursuit of a lasting peace rather than turn inward, away from the rest of humanity.

The previous year, in 1920, the League of Nations was founded in an effort to bring the nations of the world together to foster a global environment conducive to peace. In doing so, it established a number of committees to begin the painstaking work necessary to make peace an enduring reality. Part of this effort included the creation of an International Committee on Intellectual Cooperation, a group dedicated to "promoting intellectual work and international relationships" between members of intellectual professions.[15] Notable in its formation was the specific guideline that the committee should include women among its ranks, leading to Marie Curie being nominated to become part of this effort. In keeping with both of their desires to construct a more harmonious world, Curie as well as Einstein began service on this committee. Einstein was initially hesitant due to the lack of the League's complete dedication to pacifism, but he eventually relented, finally joining the committee four years after its formation but participating only sporadically.

Curie was almost as devoted to the committee's efforts to promote international cooperation in the sciences and education as she was to her own continuing scientific pursuits. She had routinely declined to participate in a host of other organizations, but her daughter Eve noted that her service on this committee "was to be her only infidelity to scientific research."[16] Curie was a major power behind the committee's projects that strove to develop protections for intellectual property rights across the globe, to support and review international scholarship and research grant funding, and to generally educate schoolchildren on the mission and activities of the League.[17]

Although they invariably agreed on most actions to be taken on individual projects, usually with Einstein following Curie's lead, typically it was Einstein who garnered most of the headlines related to their committee's efforts. Curie was only too familiar with playing second fiddle to Einstein's overshadowing presence that might attract ultimate attention for their work. She was by now grudgingly accustomed to the secondary position society still deemed suitable for women, even those rare few of her scientific stature. In any event, she inherently shunned the spotlight, secure in her self-knowledge that she had worked with a committed passion to make these projects viable and ultimately successful, which in the end was all that really mattered to her.

Certainly, Curie and Einstein had other thoughts in common as well. They had first become acquainted at the 1911 Solvay Conference, followed by Einstein's initial letter of unwavering support to Curie within weeks of her return from the conference to Paris, and the hailstorm of public criticism over her affair with Langevin. During the summer of 1913, Curie, her two girls, and their governess arranged to meet Einstein and his young son Hans Albert on a hiking vacation in the Swiss Alps. Curie had spent much of the previous year recovering from physical illness as well as mental exhaustion after her ultimately disastrous liaison with Langevin. The summer holiday would be remembered by both scientists as a pleasant interlude for the families, a time when the respect of each scientist for the other deepened. As her daughter Eve described it, "They admired each other . . . Marie, with her exceptional mathematical culture, was one of the rare persons in Europe to understand [Einstein]."[18]

Einstein had always enjoyed the fresh air and natural beauty of walking a trail in the Alps. It often allowed him uncluttered time to think, an even more favorite pastime. Curie, newly recovered from her ordeals, felt the outdoor holiday might be good for herself as well as her daughters. The families joined a few other hikers in the Alpine passes of southeastern Switzerland, near the Italian border.[19] The air was invigorating and the forested trails refreshingly cool beneath the boughs of the trees. Many of the crystal-clear views were nothing short of spectacular, with snow-tipped mountains and deep glacial lakes appearing seemingly at almost every turn. At one observation point along the trail, Curie challenged Einstein to name the Alpine peaks dotting the magnificent panorama.[20] More serious topics of discussion included their theoretical scientific musings of the moment. The youngsters, bored when hearing their elders delving into the mysteries of the universe, either scouted ahead or straggled behind, coming together once again as the party took a break from hiking to rest or for a meal.

While their shared vacation time was certainly congenial, Einstein's opinion of Curie the person appeared to diverge noticeably from his respect for her as a scientific peer. He remarked to his future wife Elsa in a letter after the vacation that Curie was a "Häringseele," literally translated as "having the soul of a herring," or, more colloquially, as "cold as a fish." Thus, part of his letter to Elsa read, "Madame Curie is very intelligent, but is as cold as a fish, meaning she is lacking in all feelings of joy or sorrow. Almost the only way in which she expresses her feeling is to rail at things she does not like."[21] Some softened the translation to equate to "meager in emotion."[22] Either way, Einstein's comment showed little understanding of Curie's delicate emotional condition. Her personal plight, beginning years before with the tragic death of her young husband and culminating more recently with the collapse of her relationship with Langevin, while accompanied by her lifelong struggles for women's rights in a scientific community wholly dominated by men, could easily be seen as the basis of this hardening effect on her emotions. It no doubt produced a harsh outlook on life. Consequently, her already intense persona was tinged with disappointment, leaving little room for experiencing many of the simple joys of life that came more easily to Einstein.

Curie was inherently imbued with a singular work focus that was relentless in pursuit of any goal. She never spared physical efforts that sorely taxed her stamina to achieve what her tremendous spark of intellect told her was attainable. This enduring trait was on display dating back to her days at the Sorbonne, with Marie's relentless dedication to education stretching her days long into each night. It evidenced itself again in the Curies' austere laboratories in Paris where she toiled with Pierre for years to separate, distill, and refine tons of pitchblende ore residue to recover fractions of grams of pure radium salts. Subsequently, this same internal fire prompted her creation of mobile X-ray units to take to the front in World War I to assist surgeons in their efforts and ultimately save lives. Once she had conceived the idea of these portable machines, Curie had to equip vehicles to bring them to the battlefield as well as train people in their use, challenges she surmounted with her indefatigable nature and will of steel. Her fierce dedication to her perceived mission and the white-hot intensity of effort required to achieve it could only come from a place deep within that merged the genesis of an idea with absolute devotion to its realization. This was, for Curie, the soul of her genius. And all the outward distractions that life threw in her path were only obstacles that needed to be overcome on the way to supreme achievement.

Einstein would become more cognizant of this as time passed and his singular vision to develop a unifying theory of physics continued to lurk frustratingly beyond his grasp. At the very end of his life, still driven by his elusive search, found next to Einstein's deathbed were scribblings of equations that he hoped would lead to the solution of his perhaps improbable unifying hypothesis.[23] It had been a puzzle that had left him searching for the missing pieces that he felt were within reach if only he worked hard enough to discover them. But he never did.

Both of these individuals were initially able to catch the glimmering of each other's brilliance at the First Solvay Conference on Physics in 1911. The gathering itself would eventually be viewed as an elaborate process experiment. Conceived by a scientist searching for a means to validate his own work in the hopes of winning a Nobel Prize for himself, sponsored by a millionaire businessman with his own scientific hypotheses, it had sought answers by bringing together some of the most intelligent people in the world to mull over the chaotic state of classical physics and its fraying explanations of how the universe worked.

The relatively small group spent less than a week presenting papers to each other and discussing the meanings contained in the presentations. All were made aware of the current state of affairs related to a view of the world from the atomic perspective. No unanimous agreements on the pressing issues of the day emerged. But the effort was enough to convince Ernest Solvay to fund the following spring what was termed the International Institute of Physics. Rather than Nobel Prizes for accomplished investigations, the Institute would support grants for worthy experiments on future work, stimulating research on an international level. Hendrik Lorentz, the masterful facilitator of the First Conference, would chair the committee to select the appropriate experimentation to be funded.

The Solvay Conference on Physics would gather once more, in 1913, before the effort was interrupted by World War I the following year, then move to a triennial cycle beginning in 1921. After the war, the Solvay Conferences were instrumental in helping to slowly bring German scientists back from an ostracized existence into the mainstream of the scientific community. In fact, Einstein was the only German scientist invited to the first Solvay Conference to convene once the war had ended, the Third Solvay Conference on Physics in 1921. The French contingent, led by Marie Curie and Paul Langevin, still polite friends after all they had been through, really didn't consider Einstein a German at all, dating back to his Swiss patent days and his Prague professorship. Although he didn't attend due to his prior commitment to make his initial visit to America, he was once again back in the fold by the Fifth Solvay Conference on Physics in 1927, along with other German physicists after their country joined the League of Nations in 1926. The Fifth Solvay Conference was to become its most famous, simply because almost 60 percent of its twenty-nine attendees were to receive Nobel Prizes. The running argument on subatomic probability versus certainty between Niels Bohr and Albert Einstein that began at this gathering was almost as notable among the world of physicists.

Ernest Solvay had passed away by then, dying in 1922 at the advanced age of eighty-five. He had lived to see the establishment of the ongoing conferences and institutes that he funded in dedication to his thirst for scientific knowledge. He had spent much of the second half of his life grappling with the mysteries of physics, chemistry, physiology, and sociology, hoping to uncover a unifying principle of his own that could help these disciplines improve society. Although unsuccessful in this regard, he ultimately directed his passion toward the use of his fortune and his networking expertise to the development of an organizational mechanism that created a unique forum for Curie, Einstein, and other geniuses of the world to debate and explore the mysteries of the scientific universe on a professional basis as well as to foster personal relationships among its attendees. This format, along with philanthropic funding for experimental investigations, was instrumental in changing the course of scientific discourse and exploration.

Marie Curie and Albert Einstein were among the most extraordinary of these scientists, brought together at the First Solvay Conference on Physics in 1911 to contemplate the radical changes taking place in the world of physics. Curie was already the face of women in science; Einstein was to become synonymous with genius. Those in attendance at the First Solvay Conference roundly came to the conclusion that Einstein was a shining star to which physics should hitch its future. The conference laid the groundwork for his ever-ascending trajectory toward worldwide fame. For Curie, it was a bit of a different story. The gathering allowed many a glimpse of the woman behind the well-known intense visage, someone with hopes and dreams like anyone else, whether in attaining another Nobel Prize or fiercely defending her right to love again.

EPILOGUE

Follow the Science

By the beginning of the 20th century, the world was awash in invisible scientific phenomena. Decades before, by the mid-1860s, unseen electric and magnetic fields and their interactions had already been shown to exist by the theoretical work of Michael Faraday and James Clerk Maxwell. Experimental physicists confirmed these theories in the 1880s. Cathode rays, composed of streams of electrons, were being investigated by many physicists, leading to Röntgen's detection of X-rays. Becquerel rays from uranium had been discovered shortly after, which begat the Curies' subsequent exploration of radioactivity, of which alpha, beta, and gamma rays were subclasses based on the particle composition and electric charge each carried. Then, in 1903, the year in which the Curies were given their Nobel Prize in physics, came the announcement of the N-ray.

René Blondlot, a prominent physicist at the University of Nancy in northeastern France, had made an unexpected and quite perplexing discovery. While experimenting with X-rays, he believed he had inadvertently discovered a new type of electromagnetic radiation, which he called N-rays after his city and university. In his experimentation, he stated that although the rays produced "no direct photographic effects," as would X-rays, they could be demonstrated by the increased intensities of barely detectable sparks when the rays were projected upon them.[1] This was a rather controversial approach to verifying the theory, relative brightness of a spark being difficult to measure and subject to the interpretation of each individual viewing the spark. Looking back on the situation, the historian M. J. Nye wrote in the chronicling of this event, "Oblivious or indifferent to the very complicated physics of variations in intensity of an electric spark in the vicinity of a cathode tube, Blondlot pursued his announced discovery. The perils of using as his primary data the extremely subjective variations in brightness of a feeble electric spark did not daunt him in the least."[2] Blondlot's experiments continued, with his instructions to those viewing the sparks that they should not actively try to put effort into seeing them.[3] Those not able to adequately detect the varying intensities of the sparks obviously did not have the required visual acuity to do so, according to Blondlot.

Experimentation continued into 1904, with Blondlot and others at the University of Nancy and across Europe trying their hand at producing N-rays. Within a very short time, some physicists claimed that many sources of the rays existed, supposedly detecting their emanation from the sun, certain metals, and even the human nervous system, internal muscles, and organs.[4] There was a thought that N-rays, like X-rays with their associated pictures of bones within the body, could be used to image internal organs for diagnostic purposes.

Some physicists said they could see a spark's increased intensity when demonstrations were given by Blondlot. A few actually described experiments that confirmed his discovery, including Jean Becquerel, son of one of the discoverers of radioactivity, Henri Becquerel. But most expressed their inability to reproduce the phenomenon, even when attempting to follow Blondlot's general experimental protocols. In France, a number of brilliant physicists including Pierre Curie, Jean Perrin, and Paul Langevin were disappointingly unable to detect N-rays when they tried to investigate the situation. Moreover, well-known scientists such as Lord Kelvin of England and Germany's Heinrich Rubens could not produce N-rays after days or weeks of effort and were developing a healthy skepticism of the whole topic.

Finally, Robert W. Wood, a respected American physicist and optics expert from Johns Hopkins University, after failing to reproduce Blondlot's mysterious rays for himself, was asked by a few of his peers to travel to France in mid-1904 to hopefully see it all firsthand. As reported in an article Wood wrote for the journal *Nature* shortly after his visit, he arrived at Blondlot's laboratory for what was to be a convincing series of demonstrations of N-rays but left "with a very firm conviction that the few experimenters who have obtained positive results have been in some way deluded."[5] Wood witnessed a number of experiments that were, to his way of thinking, subjective at best and deceptions at the worst.

All of the experiments took place in a blackened room in order to increase the visibility of the intensities of sparks that were the supposed effect of N-rays. Wood was intent on disproving the N-ray concept if given the chance. Under cover of the room's darkness, he surreptitiously adjusted or removed what had been deemed by Blondlot key pieces of indispensable equipment for the successful completion of the experiments. The results were no noticeable effects on the outcomes, according to the unsuspecting Blondlot.

In one particular instance, Blondlot set up an experiment for Wood's viewing that centered on an aluminum prism that was ostensibly used to separate the N-rays into a spectrum. The resulting prism's separation of the rays was to yield an increase in brightness along points of a phosphorescent strip on which the rays shone. Although Woods couldn't see any areas of increased brightness on the strip, his host clearly stated there was such an effect. Wood then asked the scientist to repeat the experiment and note the observed points of brightness again. "Unseen by Blondlot and his assistant, Wood removed the crucial prism from the apparatus . . . Not knowing the prism had been removed, Blondlot continued to insist he saw the same pattern he had claimed to see when the prism was in place."[6] After a few more of these experiments were discreetly altered by Wood with his covert actions, with no apparent effect on the outcomes, he was convinced of the total lack of proof of the existence of N-rays. He wrote in his article, "I am obliged to confess that I left the laboratory with a distinct feeling of depression, not only having failed to see a single experiment of a convincing nature, but with the almost certain conviction that all the changes in the luminosity or distinctness of sparks . . . are purely imaginary."[7]

The publishing of Wood's piece did prove one thing, the quick demise of the hypothesis of the N-ray within most scientific circles. Blondlot continued to try to demonstrate their existence sporadically over the next few years, but to no avail. For whatever reason, perhaps French nationalistic or university pride, N-rays were initially accepted as a potential new source of electromagnetic radiation. A few of those expressing support for Blondlot and his N-rays were professorial alumni from the University of Nancy, including Henri Poincaré and Marcel Brillouin, both of whom couldn't attest to the phenomenon but suggested there was no reason *not* to believe it. Coincidentally, even Ernest Solvay was tangentially connected to this scientific episode, not in having an opinion about the issue but by his trademark philanthropic efforts of having previously donated a substantial sum to the university for the building housing the Institut Electrotechnique at the University of Nancy.[8]

The plain fact was that the scientific method, a meticulous process dating back to Newton, had clearly not been uniformly applied in the case of N-rays. This entailed systematic experimentation to repeatedly produce a result that was objectively observable and measurable. The methodology had served science well over the years. It was demonstrated yet again in the early 1900s with the Curies' rigorous investigation of radioactivity and its subsequent support from experimental stalwarts such as Rutherford and Soddy. Einstein's general theory of relativity could not definitively be proven until Eddington returned from his solar eclipse expedition of 1919 with objective photographic evidence in hand, inherent in the use of the scientific method in order to verify Einstein's hypothesis and mathematical calculations.

Two years later, in 1921, Einstein won the Nobel Prize "for his services to Theoretical Physics." Interestingly, it had not been presented for his general theory of relativity proven by the bending of light as influenced by the sun, nor for the special theory of relativity and its associated formula of $E=mc^2$. Rather, it was awarded for his elucidation of light's influence on electrons of a metallic surface upon which it was shown, termed the photoelectric effect, and its association with energy quanta. By then receipt of his Nobel was almost superfluous, a formality in honoring someone whom the world already believed to be the greatest scientist of all time. He subsequently escaped Germany between the two world wars, settling in America and supplanting Newton in scientific grandeur based on his special and general theories of relativity as well as his work on light quanta. His second wife, cousin Elsa, whom he had married after divorcing Mileva Marić in 1919, passed away in the mid-1930s. He never remarried.

Einstein spent the remaining two decades until his death lobbying for a global government that could deliver world peace as well as attempting to uncover a unifying physics theory to connect all natural field forces. At the same time, he steadfastly supported the establishment of a Jewish homeland. The first two goals were rather lofty, needless to say neither of which were ever accomplished. But the third was more attainable and he worked as hard for that as anything else. Einstein was Jewish by birth but not religious. He had come to embrace his hereditary cultural bond with the Jewish people, hardened through the crucible of the anti-Semitism he had witnessed firsthand in Germany between the world wars and sealed with the horrors of the Holocaust.

At the same time, his two children had very different lives in store for them. Older son Hans Albert emigrated to America with his wife in the late 1930s and eventually became a well-respected hydraulic engineering professor. The younger boy Eduard, whom Albert had thought was most like himself in intelligence and manner, was to become mentally unstable and slip into schizophrenia by his twenties, being hospitalized for the bulk of his adult life in an institution in Switzerland. After Einstein left Germany for good in the early 1930s, he never saw Eduard again.

He had kept up a correspondence with both children as they were growing up, and more pointedly after his divorce from Mileva, taking an interest in each and attempting to see them when his schedule permitted. But Einstein and Mileva struggled to remain cordial, and their continued antagonism negatively impacted the relationship the sons had with the father, especially for Hans Albert. Letters, no matter how tender or understanding, were no substitute for being a father who was at their sides in their formative years, and they could no doubt see the hurt that divorce had wrought upon their mother.

As Hitler's Germany became a more violent presence in Europe, with increasing intolerance of Jews at home, Einstein encouraged his son to leave for America. In 1938, Hans Albert and his wife finally followed his advice. Initially, they moved to South Carolina for a few years where he worked for the U.S. government as a research engineer focused on riverbed sedimentation. With his own wife now deceased, Einstein began to see his son and family periodically in South Carolina, a pleasant train ride from his Princeton, New Jersey, home. The chilly interaction between father and son began to thaw; Einstein began to pick up the pieces of his badly fractured relationship with at least one of his boys. Then, Hans Albert's job moved him across the country to Caltech, after which he became a professor of hydraulics at the University of California at Berkeley, where he was a well-known expert in his field. He had found a niche in which he excelled, very separate from his father's celebrity, which suited him just fine.

Both sons had inherited a love of music from Einstein. Hans Albert was proficient on the piano, but Eduard appeared to be more gifted and decidedly more passionate about it, unfortunately playing to the point of frenzy as he grew older and his illness manifested itself.[9] Still, what musical moments father and sons might have created through the youths' early years if things had worked out differently, one or the other of the boys playing the piano in concert with their father's loving violin rendition of a Mozart sonata. With it perhaps a component of a happier family life might have been built, but it wasn't to be.

By contrast, Marie Curie remained close with her two daughters, even though her workmanlike devotion to her research sometimes made her a less approachable figure than her girls would have desired. As it turned out, Irène was to live a life that followed in her mother's footsteps after their shared experience delivering X-ray technology to the front lines during the war. She researched radioactivity with Marie at her Radium Institute in Paris and eventually won worldwide fame and a Nobel Prize of her own in 1935. This was for the discovery of artificial radioactivity, the use of radiation to create new radioactive elements from previously stable ones. Besides Marie, she had become only the second woman to be awarded a Nobel Prize. As with her mother, Irène shared the honor with her husband, Frédéric Joliot-Curie, a fellow physicist who had joined the Radium Institute after the war on the recommendation of Paul Langevin. Through the years since the dissolution of their romantic relationship, Langevin had kept on relatively good terms with Marie, who was not one to hold a grudge. Irène and Frédéric's devotion to what had become the family's radioactive destiny followed similar paths not only in worldwide recognition, but in tragic early deaths for both, Irène at fifty-nine and then Frédéric two years later, being only fifty-eight years old. Both had succumbed to radiation exposure, what Frédéric termed their "occupational disease."[10]

It appears the relatively small number of French scientists and their families were indeed as close as ever, the elite professors and researchers continuing to teach at the same top French universities, vacation at the same seaside resorts, and associate very much with each other. As biographer Barbara Goldsmith stated, "In the small, tightly knit scientific community in France it was not considered unusual that Hélène [Irène's daughter] married Michael Langevin, the grandson of Paul Langevin."[11] Of course, it was also only natural that both would become esteemed scientists in their chosen field of nuclear physics. Indeed, the world of Curie and Langevin had come full circle over the course of two generations, Hélène and Michael forming the union that had been denied their grandparents those many years before. Their son, Yves, is an astrophysicist who continues to this day the family's legacy of scientific discovery.

Curie's youngest daughter, Eve, was to add to family Nobel Prize lore, but along a more circuitous path. Unlike her mother or sister, Eve's brilliance expressed itself in the arts rather than the sciences. She was an accomplished pianist, talented enough to give concerts across France and Belgium. She eventually became a journalist during the Second World War. Marie passed way in 1934, and both daughters had diligently cared for her during her final years. By 1937, Eve had authored an international best-selling biography of Marie Curie called *Madame Curie*. The formal title at once expressed the respect she held for her mother, while the contents poured forth the undying love for Marie that resided in her heart. Eventually, Eve married an American, Henry Richardson Labouisse Jr., head of the United Nations organization UNICEF (United Nations International Children's Emergency Fund). She traveled the globe with him as UNICEF ambassadors while the organization they represented administered aid to the world's children. In 1965, Henry accepted the Nobel Peace Prize on behalf of UNICEF. The honor was given to recognize the organization's promotion of brotherhood among global nations through its work with children. This award became the fifth Nobel Prize associated with the family, after Pierre and Marie each won in 1903, followed by Marie's second in 1911, and Irène's in 1935. No family has come close to attaining such a record of Nobel success, and in all likelihood, none ever will.

Marie and Irène each died at a relatively young age, sixty-seven and fifty-nine, respectively. Both had been exposed to radioactivity and X-rays for much of their professional lives, directly leading to their demise, Marie of aplastic anemia and Irène of leukemia. By contrast, Eve, never having been a part of the "family occupation," was to live past her hundredth birthday, passing away at the age of 102 in 2007. It could be speculated that living divorced from radiation had added another thirty to forty years to her lifespan.

When Curie passed away in 1934, with the storm clouds of the Second World War forming, European scientists were becoming increasingly fearful of potential German repression. Against this backdrop, a year after Marie Curie's death, Einstein spoke in tribute to his friend at a Curie Memorial Celebration held in New York City. By then he had come to realize more completely her towering gifts to science as driven by her unmatched internal fortitude and focused his comments in part on her intensity in pursuit of scientific truth: "Her strength, her purity of will, her austerity toward herself, her objectivity, her incorruptible judgment—all these were of a kind seldom joined in a single individual. . . . The greatest scientific deed of her life—proving the existence of radioactive elements and isolating them—owes its accomplishment not merely to bold intuition but to a devotion and tenacity in execution under the most extreme hardships imaginable . . . If but a small part of Mme Curie's strength of character and devotion were alive in Europe's intellectuals, Europe would face a brighter future."[12]

The Curie women were nothing if not resolute in pursuit of their goals. With this persistence resulting in their Nobel Prize awards, Marie and Irène had appeared to pierce the barrier that existed for women's professional recognition in the sciences. But it took twenty-nine more years from the time Irène won in 1935 for another woman to be honored with a Nobel Prize in chemistry. Englishwoman Dorothy Hodges achieved the award in 1964 for her use of X-ray imaging to identify important biological substances. Forty-five more years would pass before the chemistry prize would be bestowed upon another woman, this time in 2009 to Ada Yonath for her study of the structure and function of the ribosome. More recently, in 2018, American Frances Arnold took home the prize for her work on enzymes. All told, in almost 110 years since Marie Curie had become the first woman to win the Nobel Prize in chemistry in 1911, only five female scientists had managed to receive the award.[13]

The numbers had been even stingier on the physics side of the Nobel ledger. After Marie shared the 1903 prize with her husband and Henri Becquerel, sixty years went by before Maria Goeppert Mayer was to win the second physics award for a woman when she was recognized in 1963 for her work on nuclear shell structure. Another fifty-five years passed before Donna Strickland won in 2018 for groundbreaking inventions in the field of laser physics, making three women physics Nobel laureates since 1903.[14] All told, eight Nobel Prizes had been awarded to women across the scientific disciplines of chemistry and physics in the 118 years since the awards began in 1901. Someone checking the history would be forgiven for thinking there must have been some sort of clerical error in the counting, but the numbers didn't lie. At the same time, over that period a dozen women had won Nobel Prizes in the category of physiology or medicine, sixteen in literature, seventeen for their efforts to bring peace. Often until recently, the latter two categories had sometimes been whispered as being "more suitable" to women's temperament and characteristics, this notion appearing to reflect the remnants of mid-19th-century thinking still stubbornly embedded in the shadows.

Then, as the announcements for the 2020 Nobel Prize awards were made public during the week of October 5, ripples of excitement began to course through the scientific world. On Tuesday, October 6, Andrea Ghez, an astrophysicist from UCLA, was awarded a portion of the physics prize for her work on supermassive black hole investigation and discovery. The following day, Jennifer Doudna and Emmanuelle Charpentier captured the chemistry award for their joint efforts in the development of a unique method for genome editing. These three women had succeeded in leaping past the seemingly insurmountable obstacles preventing recognition for female scientific efforts. The trio represented the combined equivalent of almost 40 percent of the number of women who had previously received these awards from 1901 through 2019. Of course, the number of women in STEM fields (science, technology, engineering, and mathematics) had been vastly underrepresented as a portion of the female population during the 20th century and this remains so even today. In 2010, women comprised only 28 percent of those filling the scientific occupations in all science and engineering arenas globally, up from the low 20 percent range of the mid-1990s.[15] These three 2020 female awardees, coupled with the two women Nobel winners in physics and chemistry in 2018, might appear to be a signal that recognition of women's roles in scientific research is beginning to become more appreciated, perhaps resulting in enticing more women into these disciplines in the future.

Pierre Curie would have heartily supported this long-anticipated development. He had fought the battle back in 1903 when he indicated to the Nobel Prize Committee that he would rather forgo the prize in physics than unjustly deprive his wife of her well-earned portion of the award. As both he and his wife were clearly aware, intelligence knows no bounds, nor is it the province of gender or race.

The scientific genius displayed by those who attended the First Solvay Conference on Physics in 1911 was focused on one goal: to understand the intricacies of the world around them, to discuss and debate together the strange phenomena they were encountering in their investigations of the atomic world, so that they might use their intellect to see more deeply into these puzzles. Their desire was to uncover new truths and present them to a world that could employ them to move forward armed with new knowledge. Their brilliance was a vehicle to this end, to be used diligently to measure reality and use these observations to make sense of the unexplainable. Marie Curie was invited to attend this council precisely because her continuing exhibition of genius fit the mold of what was called for at the gathering, completely independent of gender.

Developing explanations for perplexing natural phenomena required those attending the First Solvay Conference to challenge deeply held theories while opening their minds to a wider range of possibilities than many others were prepared to do. The very idea of the quantum, a discrete packet of energy that represented a drastically different concept than the classical physics of energy continuity held so dear by the scientists at the conference, was enough to trouble many of the participants. As the physicist and historian Gerald Holton recounted, "Nernst . . . had said that quantum physics was at bottom 'a very odd rule, a grotesque one.' Max Planck wrote afterward, 'For my part, I hate discontinuity . . .'"[16] This, from the man whose determined investigations of heat phenomena had led him to the very mathematical construct of the energy quantum in order to adequately explain his observations. But, as Holton had noted, thematic scientific tenets ingrained in these physicists, though serving as a sound basis for their exploration of the natural world, "have to stand the test of experience and be judged by the degree to which they contribute to making the world of phenomena more 'intelligible.' Nature cannot be fooled."[17]

High upon the northeastern ridge of the Matterhorn, the mountain climber's iconic, natural pyramidal peak standing over 14,700 feet in the Swiss Alps, is a small structure originally placed there in 1915. No more than a glorified hut balanced precariously on a tiny outcropping of the severely sloping shoulder of the trail, it is a last refuge for weary adventurers in emergency need of shelter from treacherous snow and ice, driving rain, or forceful winds that can often suddenly appear with an alarming frequency while journeying up or down the mountain.

The cabin was constructed from materials hoisted up the daunting mountainside by cable from camps farther below. Over the years, parts of the structure have been rebuilt, and it has tight sleeping quarters that can accommodate up to ten. In most instances, the presence of the hut is a convenient marker for experienced climbers to indicate they have covered about three-quarters of the hike to the top. But whether in signifying a point along the trail or serving the infinitely more vital purpose of providing potentially much-needed shelter against the sometimes unruly but ever-present elements of nature, the Solvay Hut, as it is known, is always a welcome sight.

At its inception, fifty years after the first successful scaling of the Matterhorn in 1865, the construction had been financed by a donation from the experienced climber Ernest Solvay, who as an Alpine mountaineering enthusiast would continue to actively pursue this pastime until his early eighties.[18] Solvay had survived a bout with pleurisy in his youth, keeping him homebound and robbing him of his dream to go to college and become a degreed scientist. But, just as he overcame the physical setback of his late teens by becoming an avid climber in the Alps, he was to surmount his lack of formal training as a university-bred scientist to establish one of the greatest chemical enterprises in the world, his fierce determination to succeed the basis of his efforts both personally as well as professionally.

Solvay had donated millions of francs from his own fortune to numerous scientific causes, most important of which were to sponsor the conferences and institutes that bore his name. For him, a comparatively modest 20,000 francs to build the hut, though not a contribution to science, was perhaps as important to his Alpine climbing passion. Reaching the building on the ascent signified a climber's determined efforts to make it to this point in the dangerous and exacting adventure, the goal of attaining the summit now within reach. The rigorous planning and preparation for the trip, both mental and physical, were part of the total commitment required for the journey. Without it, the success of the mission couldn't be achieved. Yet, sometimes circumstances beyond a climber's control, poor weather or physical injury, dictated the ascent might all be for naught. It was in these extreme instances, that only a seasoned mountain climber had either experienced or anticipated, where Solvay wished to offer help through the presence of his tiny shelter at a crucial point in the journey. It represented a sign of hope and assistance, perhaps even propelling to the top the determined Alpinist faced with adversity.

Similarly, Solvay was dedicated to supporting his exclusive scientific meetings that had turned into iconic gatherings in their own right. As well, he established his scientific institutions that sponsored creative investigation of natural phenomena. These were signature efforts in his plan to employ science to assist the world in understanding its most complex problems, to help propel it into an uncertain future. Only an elite few with the commitment of furthering exploration to the very edges of understanding the physical and chemical workings of the world were invited to participate in his meetings or were awarded highly selective research grants. Only those worthy of the climb invariably got involved.

Solvay had considered the First Solvay Conference on Physics of 1911 a success, if not in solving the conundrums that faced the world of atomic physics at this crucial point in time, then at least in laying them bare for future investigation and resolution. His next step was to collaborate with the conference facilitator, the universally respected Dutch scientist Hendrik Lorentz, to develop the International Solvay Institute of Physics the following year to support further explorations of these issues. Similarly, an International Institute of Chemistry was founded in 1913, and the First Solvay Conference on Chemistry was established in 1922. Approximately every three years, the conferences have brought together a limited number of the leading-edge scientific thinkers for a week of examination into pressing problems in their discipline. The Institutes have supported what they term "curiosity-driven research" to investigate promising solutions to these issues.[19]

Over the course of the past 110 years, only suspended for the global turbulence of two world wars, the Solvay Conferences have managed to withstand the test of time. In 2020, due to the coronavirus pandemic, the Twenty-Eighth Solvay Conference on Physics, originally scheduled for October, was understandably postponed until October 2021. Somehow, it seemed inappropriate to substitute a week of Internet meetings for what had always been an intimate, small group conference that placed a premium on in-person interaction and idea exchange.

Jean-Marie Solvay, the great-great-grandson of Ernest, heads the Solvay Institutes and is responsible for the continuation of the Solvay Conferences, as well as having a long involvement on the Solvay company's board of directors. In a recent conversation to which Jean-Marie graciously agreed, he commented from the vantage point of a thoughtful, resolute business veteran with deeply held opinions about the need for the continuation of the scientific conferences and institutes as well as the future of the corporation established by his ancestor and still largely owned by the Solvay family.

On the subject of the company, he commented, "Solvay needs to maintain and grow its place in the chemical industry by being agile in meeting customers' needs, entrepreneurial in its approach to solving problems, and tenacious in achieving its goals."[20] When asked how to do all of this, his simple reply might have come from his great-great grandfather, "It's all about the quality of the people you employ."[21] Ernest Solvay had great trust in the networks he gathered around him to further his scientific investigations as well as to help devise and execute the business plans he put into motion. He desired to find the brightest minds in physics and chemistry to bring to his conferences, institutes, and personal scientific investigations. In his business, he wanted the same determination that he brought to the enterprise, whether from his management team or down to the daily laborers in his business factories, whom he knew would give their all if treated fairly.

Related to the conferences, Jean-Marie had previously noted, "The concentration of energy, passion, and intensity on display is fundamentally driven by the scientists' curiosity about how the world works. It's an intense human adventure."[22] In our discussion, he now suggested that the biggest challenge for the institutes and conferences was "staying relevant in order to have the largest impact on the future. These ventures are instrumental in influencing science, from the standpoints of helping to determine what the biggest scientific questions are that face the world, who are those most equipped to discuss these topics, and how best to fund experimentation to advance learning about how to solve these challenges. We choose the subjects of the conferences carefully, avoiding the influence of trends, to focus on the critical few areas that require the most problem-solving thought. And we must continue to identify scientific talent of the highest caliber to address these issues."[23]

Marie Curie and Albert Einstein were two such uniquely talented individuals. They investigated mysteries of nature somewhat as they might approach a hike in the mountains, a favorite pastime of both. Careful preparation was imperative and detailed observation of the sights along the way was required, driven by a desire to expand on an understanding of the natural phenomena being investigated. They never stopped imagining or appreciating the variations and dimensions in which the world could exist as they observed their surroundings. In scientific practice, such effort yielded the amazing discovery that atomic radiation was caused by degradation of an element itself. It generated the startling realization that light in the universe bent around the sun due to that huge star's warping of space-time. These hypotheses were proven by repeatable experimentation over time. If they couldn't be verified in this manner, then the theories needed to be reexamined to better understand the phenomena in greater detail.

As we have found in today's most challenging of times, sometimes it's hard to believe that what we are observing of the world around us is actually happening. We are in the midst of a global pandemic the likes of which mankind hasn't experienced for a hundred years, over 100,000,000 infected, millions dead. The world's climate events are becoming more severe each day. Hurricanes are more numerous and deliver greater devastation, and polar icecaps and glaciers are quickly melting, raising sea levels and endangering coastal inhabitants. In North America, record numbers of fires are raging out of control for months on end, swiftly moving infernos that we must spend increasingly more time and effort trying to control while human life hangs in the balance. And each year, the world grapples with determining the right approach to keeping clean the very air we breathe and the water that surrounds us. While we dally in agreeing on a decisive plan to alleviate the situation, each of these most valuable of natural resources is becoming increasingly contaminated, used as a dumping ground for the excesses of our continued industrialization.

How do we address these issues? One might suspect that Ernest Solvay would have suggested that we investigate these complex problems by employing the best tools available to us—people with deep understanding and experience in the fields in which they have extensive training. That we encourage these individuals to build on the ideas of those who have gone before, as Isaac Newton himself described his discoveries in a letter to fellow scientist Robert Hooke, "if I have seen further, it is by standing on the shoulders of giants."[24] That we provide these people with the most innovative means that we can afford to give them for their efforts, and have them use these instruments and processes to take continuous, detailed measurements of the phenomena they are evaluating and engage in in-depth discussions about their findings. Then we encourage them to provide their analyses of, and their thoughts on, what they have observed. We need to examine their theories and see if there exists general agreement among these individuals on their findings, providing greater confidence to rely on the ideas being presented to us.

One important means of supporting the efforts of these scientists is the maintenance of networking opportunities and experimentation funding that the Solvay Conferences and Institutes and Nobel Prizes continually offer for the advancement of scientific discovery. But there are many other options that we can choose to employ, including government involvement and assistance, private industry focus and participation, and even something as simple as the public's expression of its belief in the worth of science in solving the complex issues that impact our future.

Undoubtedly, some theories that will be proposed by the scientists will be shown to be invalid through rigorous questioning and examination employing the scientific method, much like the fate of the N-ray. But if no better explanations for the phenomena that have been thoroughly investigated exist, it invariably becomes time for people to use this information, to act based on these theories, which represent the best of our knowledge to date. So it was that we used the discovery of radioactivity to attack the scourge of cancer. We employed the understanding that mass and energy are equivalent to unlock the unfathomable power of nuclear fission. And we strove to guide and regulate what was being recommended to address issues and concerns raised by the theories, and as more information became available, we adjusted our actions accordingly. Through it all, we learned that we must not be held prisoner to a scientific hubris that states that there is nothing more to explore and learn, and that there are no new conclusions to be drawn from these efforts.

As we once knew it, the sun revolved around the Earth. Then, through consistent observation, evaluation of data, and the diligent application of extraordinary intelligence, the exact opposite was shown to be true. It took time for us to accept this revolutionary new theory and other ideas that followed concerning the very basis of our physical existence. However, as we began to understand the world through these new perspectives, offered for our consideration by a select few individuals with exceptional insights that transcended human imperfection, we began to make impactful decisions about our world based on this new knowledge. In doing so, we became beneficiaries of the soul of genius.

ACKNOWLEDGMENTS

This, my first effort at full-length narrative history authorship, has truly been a labor of love. From its conception to its planning, through its research and writing, to all the myriad details related to getting the manuscript in shape to publish, it's been at times an exhilarating as well as draining experience. I feel it all has come together in an unforgettably special way. To arrive at this point, it was so important to have the guidance and support of people who wanted me to succeed.

To thank everyone for all of their engagement, input, and positivity would just begin to scratch the surface of my indebtedness to each of them. From the moment I received an unsolicited email with the subject line "From an Agent," I couldn't have expected the interest Susan Canavan of Waxman Literary would show in the project I suggested we develop. Her realistic approach to the effort assisted me in crafting just the right book proposal as a new author without letting me drown in my own naiveté. This led us to the opportunity to find the perfect publishing house for the story. Jessica Case at Pegasus has been all I could have hoped for in an editor, diligent in her appraisals of my writing, generous in her praise, and skillful in her direction when I headed too close to the rocks.

I was fortunate enough in my research to discover an important piece of my narrative in the forgotten scribblings of French scientist Marcel Brillouin. These notes, penned in French, were offered to me in the digitized version of their original form by the American Institute of Physics. The staff there assisted me in accessing this information, and I thank them for making this possible. Once obtained, the laborious task of translating this cache of hand-written French into intelligible English was assumed by my friend Jean-Matthieu Hautenauve and his mother Catherine. I greatly appreciate their meticulous work. Any mistakes in the interpretation of these translations are mine.

My insights into the world of industrialist Ernest Solvay were given special focus by my contact with Solvay's Corporate Heritage Manager Nicolas Coupain. Jean-Marie Solvay's views on the company and the Solvay Conferences were an added bonus. A sincere thanks to both of them for their comments on my work, as well as for providing access to the Solvay Heritage Collection photos that appear in this book. Many thanks to Matthew Stanley as well, who took the time from his own writing and teaching schedule to review my attempts at simplifying physics concepts into layman terms.

As the writing continued, I could only proceed, a chapter at a time, by turning to my secret weapon of a coach, counselor, and cheerleader, my wife Debbie. Her background as an attorney turned high school English teacher proved the perfect foil to my wandering prose; she was incisive in commenting on my story, and gently directive when needed. And my daughter Rebecca, another lawyer with a science-fiction leaning to her reading, objectively told me if my depiction of scientific phenomena was too obtuse. Together, this family legal trust was invaluable to my efforts. And, of course, numerous friends and family members offered suggestions. Even more importantly, they graciously suffered through my monologues concerning a host of different topics and personalities from the story which I continually dragged into most every conversation. I thank them all for their patience and good wishes.

The First Solvay Conference on Physics was memorable as a gathering that brought brilliant scientific minds together to examine the mounting conflicts in physics. I'm so pleased that my first book has benefitted from having the input and guidance from my own group of advisors noted above, who helped me bring this significant historical episode to life.

ENDNOTES

Introduction

1 Albert Einstein, letter to Max Born, December 4, 1926, *The Collected Papers of Albert Einstein*, vol. 15, The Berlin Years: Writings & Correspondence, June 1925–May 1927. doc. 426, 403, http://einsteinpapers.press.princeton.edu.

Chapter One: The First Solvay Conference: A New Approach

1 Barbara Goldsmith, *Obsessive Genius: The Inner World of Marie Curie* (New York: W.W. Norton Paperback, 2005), 170.
2 Jean-Marie Dilhac, "Edouard Branly, the Coherer, and the Branly Effect," *IEEE Communications Magazine*, September 2009, 24, https://doi:10.1109/MCOM.2009.5277488.
3 Susan Quinn, *Marie Curie: A Life* (Cambridge, Mass.: Da Capo Press, 1995), 292.
4 Frits Berends and Franklin Lambert, "Einstein's Witches Sabbath: The First Solvay Council on Physics,"*Europhysics News* 42, no. 5 (2011): 15, https://www.europhysicsnews.org/articles/epn/pdf/2011/05/epn2011425p15.pdf.
5 Ernest Solvay, letter to Albert Einstein with an invitation to the Solvay Conference, June 9, 1911, *The Einstein Papers*, vol. 5, doc. 269, 190–191.
6 "Lights All Askew in the Heavens," *Special Cable to tyhe New York Times*, November 10, 1919, https://www.nytimes.com/1919/11/10/archives/lights-all-askew-in-the-heavens-men-of-science-more-or-less-agog.html.
7 Alfred Nobel, "The Will of Alfred Nobel," trans. Jeffrey Ganellen, *Nobel Media*, 2018, http://www.nobelprize.org/alfred_nobel/will/.

8 Goldsmith, *Obsessive Genius*, 109.

9 Ibid., 108.

10 Eve Curie, trans. Vincent Sheean, *Madame Curie: A Biography* (New York: Doubleday, 1937; Econo-Clad Books, 1986), 206.

11 Ibid., 207–208.

12 Ibid., 211.

13 Kenneth Bertrams, Nicholas Coupain, and Ernest Homburg, *Solvay: History of a Multinational Family Firm* (Cambridge, UK: Cambridge University Press, 2013), 138, fn 69.

14 Elisabeth Crawford, "The Solvay Councils and the Nobel Institution," *The Solvay Councils and the Birth of Modern Physics*, Pierre Marage and Gregoire Wallenborn, eds. (Basel: Birkhauser Verlag, 1999), 48.

15 Didier Devriese and Gregoire Wallenborn, "Ernest Solvay: The System, the Law, and the Council," *The Solvay Councils and Birth of Modern Physics*, Marage and Wallenborn, eds., 3.

16 Berends and Lambert, *Einstein's Witches Sabbath*, 16.

17 Stéphane Foucart; trans. Crispin Gardiner, "At the Métropole, a 'Witches Sabbath'" *Le Monde*, July 31, 2015, http://quantum.otago.ac.nz/download /solvay.pdf.

18 Einstein, letter to Walther Nernst, June 20, 1911, *The Einstein Papers*, vol. 5, doc. 270, 192.

19 Walther Nernst, letter to A. Schuster, March 16, 1910, Royal Society, London, quoted in Diana Kormos Barkan, *Walther Nernst and the Transition of Modern Physical Science* (Cambridge, UK: Cambridge University Press, 1999), 183.

20 Jagdish Mehra, *The Solvay Conferences on Physics, Aspects of the Development of Physics Since 1911* (Dordrecht, Netherlands: D. Reidel Publishing Company, 1975), xxi.

21 Albert Einstein, letter to Carl Seelig, 1952, quoted in Helen Dukas and Banesh Hoffman, eds, *Albert Einstein, The Human Side, Glimpses from His Archives* (Princeton, N.J.: Princeton University Press, 1979), 6.

22 Walter Isaacson, *Einstein, His Life and Universe* (New York: Simon & Schuster Paperback, 2008), 156.

23 Einstein, letter to Michele Besso, September 11, 1911, *The Einstein Papers*, vol. 5, doc. 283, 204.

24 Einstein, letter to Besso, October 21, 1911, *The Einstein Papers*, vol. 5, doc. 296, 214.

25 Mehra, *Solvay Conferences on Physics*, xxii.

26 Carl Seelig, *A Documentary Biography* (Zurich: Europa Verlag, 1954), 162–163, quoted in ibid., xxii.

27 Einstein, letter to Heinrich Zangger, November 15, 1911, *The Einstein Papers*, vol. 5, doc. 305, 222.
28 Einstein, letter to Zangger, November 7, 1911, *The Einstein Papers*, vol. 5, doc. 303, 219.

Chapter Two: The Intensity of Marie Curie

1 Harold B. Segel, *Renaissance Culture in Poland: The Rise in Humanism, 1470–1543* (Ithaca, N.Y.: Cornell University Press, 1989), 4–5.
2 Jan K. Ostrowski, "Art in Poland from the Renaissance to the Rococo," 21, http://archiv.ub.uni-heidelberg.de/artdok/3003/1/Ostrowski_Jan_K_Art_in_Poland_2011.pdf.
3 Ilia Rodov, *The Torah Ark of Renaissance Poland: A Jewish Revival of Classic Antiquity* (Leiden: Brill, 2013), 29.
4 Stephen P. Mizwa, "Nicolaus Copernicus," *Publications of the Astronomical Society of the Pacific* 55, no. 323 (1943): 67–68, http://jstor.org/stable/40669774.
5 Sheila Rabin, "Nicolaus Copernicus," *The Stanford Encyclopedia of Philosophy*, Fall 2019 edition, http://plato.stanford.edu/archives/fall2019/entries/copernicus.
6 Mizwa, "Nicolaus Copernicus," 65.
7 Marie Curie, trans. Charlotte and Vernon Kellogg, *Pierre Curie: With Autobiographical Notes by Marie Curie* (New York: Macmillan Company, 1923), 159.
8 Goldsmith, *Obsessive Genius*, 26.
9 M. Curie, *Pierre Curie*, 157–158.
10 Goldsmith, *Obsessive Genius*, 29.
11 E. Curie, *Madame Curie*, 41.
12 Goldsmith, *Obsessive Genius*, 36
13 Ibid., 38.
14 M. Curie, *Pierre Curie*, 167.
15 E. Curie, *Madame Curie*, 86.
16 Quinn, *Marie Curie*, 97.
17 Nobel Prize Online, "Gabriel Lippmann—Biographical," *Nobel Media*, http://nobelprize.org/prizes/physics/1908/lippmann/biographical.
18 Gerhard Heinzmann and David Stump, "Henri Poincaré," *The Stanford Encyclopedia of Philosophy* (Winter 2017), https://plato.stanford.edu/archives/win2017/entries/poincare.
19 E. Curie, *Madame Curie*, 97.
20 Ibid., 110.
21 M. Curie, *Pierre Curie*, 170–171.

22 Quinn, *Marie Curie*, 105.
23 M. Curie, *Pierre Curie*, 173.
24 E. Curie, *Madame Curie*, 121.

Chapter Three: Life Partners

1 Margalit Fox, "Margaret Abbott, 1878–1955," *New York Times*, March 8, 2018, https://www.nytimes.com/interactive/2018/obituaries/overlooked-margaret-abbott.html.
2 Mary McAuliffe, *Twilight of the Belle Epoque* (Lanham, Md.: Rowman & Littlefield, 2014), 7.
3 Olympics News, "Paris 1900: Summer Olympics—Results & Highlights," http://olympic.org/paris-1900.
4 Fox, "Margaret Abbott."
5 Rita Amaral Nunes, "Women Athletes in the Olympic Games," *Journal of Human Sport and Exercise* 14, no. 3 (2019): 676, http://doi.org/10.14198/jhse.2019.143.17.
6 Linda L. Clark, *Women and Achievement in Nineteenth-Century Europe* (Cambridge, UK: Cambridge University Press, 2008), 22.
7 Ibid.
8 Charles Darwin, *The Descent of Man, and Selection in Relation to Sex* (London: John Murray, 1871), vol. 2, 316–327, http://darwin-online.org.uk/EditorialIntroductions/Freeman_TheDescentofMan.html.
9 Ida M. Tarbell, *The Book of Woman's Power* (New York: Macmillan Company, 1911), xii–xiv.
10 Clark, *Women and Achievement*, 187.
11 Ibid., 190.
12 Ann Hibner Koblitz, "Career and Homelife in the 1880s: The Choices of Mathematician Sofia Kovalevskaia," in *Uneasy Careers and Intimate Lives, Women in Science, 1789–1979*, ed. Pnina G. Abir-Am and Dorinda Outram (New Brunswick, N.J.: Rutgers University Press, 1987), 184.
13 Joan Mason, "Hertha Ayrton (1854–1923) and the Admission of Women to the Royal Society of London," *Notes and Records of the Royal Society of London* 45, no. 2 (July 1991): 205–206, http://jstor.org/stable/531699.
14 Ibid., 209.
15 Eva Hemmungs Wirtén, *Making Marie Curie: Intellectual Property and Celebrity Culture in an Age of Information* (Chicago: University of Chicago Press, 2016), 16–17.
16 Richard Staley, *Einstein's Generation: The Origins of the Relativity Revolution* (Chicago: University of Chicago Press, 2008), 168.

17 Ibid., 168.
18 Goldsmith, *Obsessive Genius, p. 59*
19 Pierre Curie, letter to Marie Skłodovska, August 10, 1894, quoted in E. Curie, *Madame Curie*, 130.
20 Quinn, *Marie Curie*, 112.
21 Wirtén, *Making Marie Curie*, 20.
22 Goldsmith, *Obsessive Genius*, 55.
23 E. Curie, *Madame Curie*, 141.
24 M. Curie, *Pierre Curie,* 176.
25 Goldsmith, *Obsessive Genius*, 70.
26 Robert Reid, *Marie Curie* (New York: Mentor Books, 1975), 61.
27 M. Curie, *Pierre Curie*, 94.
28 Quinn, *Marie Curie*, 147
29 Goldsmith, *Obsessive Genius*, 75.
30 M. Curie, *Pierre Curie*, 97.
31 Ibid.
32 Mme. Skłodowska Curie, "Rays Emitted by the Compounds of Uranium and Thorium," April 12, 1898, French Academy of Sciences, quoted in Helena M. Pycior, "Reaping the Benefits of Collaboration While Avoiding Its Pitfalls: Marie Curie's Rise to Scientific Prominence," *Social Studies of Science* 23, no. 2 (May 1993): 304, http://jstor.org/stable/285481.
33 Ibid., 306.
34 Ibid.
35 M. Curie, *Pierre Curie*, 186–187.
36 Goldsmith, *Obsessive Genius*, p. 86.
37 Quinn, *Marie Curie*, 159.
38 E. Curie, *Madame Curie*, 176–177.
39 H. Langevin-Joliot, "Radium, Marie Curie and Modern Science," *Radiation Research* 150, no. 5, Supplement: Madame Curie's Discovery of Radium (1898): A Commemoration by Women in Radiation Sciences (November 1998), S6, http://jstor.org/stable/3579803.
40 Reid, *Marie Curie*, 95–96.
41 M. Curie, *Pierre Curie*, 133.
42 Quinn, *Marie Curie*, 206.
43 Roger F. Robison, *Mining and Selling of Radium and Uranium* (New York: Springer, 2015), 70.
44 Goldsmith, *Obsessive Genius*, 88.
45 Marie Curie, entry in "Mourning Journal," May 1906, quoted in Quinn, *Marie Curie*, 243.

Chapter Four: The First Solvay Conference: Science at a Crossroads

1 Stephen Porter, *The Great Plague* (Gloucestershire, UK: Amberley, 2009), 28–29.

2 Evelyn Lord, *The Great Plague: A People's History* (New Haven, Conn.: Yale University Press, 2014), 1.

3 Newton Additional MS. 3968.41, f.88 and f.99; Additional MS. 3996, Cambridge University Library, quoted in Milo Keynes, "The Personality of Isaac Newton," *Notes and Records of the Royal Society of London* 49, no. 1 (January 1995): 21, http://jstor.org/stable/531881.

4 H. W. Turnbull, ed., *The Correspondence of Isaac Newton* (Cambridge: Cambridge University Press, 1959), vol. 1, 59, quoted in Alan E. Shapiro, "The Evolving Structure of Newton's Theory of White Light and Color," *Isis* 71, no. 2 (June 1980): 212–213, http://jstor.org/stable/230172.

5 Keynes, "The Personality of Isaac Newton," 21.

6 Jason Socrates Bardi, *The Calculus Wars: Newton, Leibniz, and the Greatest Mathematical Clash of All Time* (Philadelphia: Perseus Books Group, 2010), 3.

7 Ethan Siegel, "When Did Isaac Newton Finally Fail?" *Forbes*, May 20, 2016, https://www.forbes.com/sites/startswithabang/2016/05/20/when-did-isaac-newton-finally-fail.

8 Richard S. Westfall, "The Career of Isaac Newton: A Scientific Life in the Seventeenth Century," *The American Scholar* 50, no. 3 (Summer 1981): 347, http://jstor.org/stable/41210741.

9 J. M. Keynes, "Newton the Man," *The Royal Society of London: Newton Tercentenary Celebrations* (Cambridge, 1947), quoted in Keynes, "The Personality of Isaac Newton," 22.

10 Nancy Forbes and Basil Mahon, *Faraday, Maxwell, and the Electromagnetic Field: How Two Men Revolutionized Physics* (Amherst, N.Y.: Prometheus Books, 2014), 17.

11 "A Few Holes to Fill," *Nature Physics* 4, no. 257 (April 2008): http://doi.org/10.1038/nphys921.

12 American Physical Society, "This Month in Physics History," *APS News* 11, no. 9 (October 2002): http://aps.org/apsnews/this-month-in-physics-history/october-2002/planck.

13 Albert Einstein, "Notes for an Autobiography," *The Saturday Review of Literature*, November 26, 1949, 11, https://archive.org/details/Einstein Autobiography.

14 A. A. Michelson, "Some of the Objects and Methods of Physical Science," *University of Chicago Quarterly Calendar*, August 1894, as quoted in Lawrence Badash, "The Completeness of Nineteenth-Century Science," *Isis* 63, no. 1 (March 1972): 52, http://jstor.org/stable/229193.

15 Alexander Bird, "Thomas Kuhn," *The Stanford Encyclopedia of Philosophy* (Winter 2018), Edward N. Zalta, ed., https://plato.stanford.edu/entries/thomas-kuhn/.

16 Diana Kormos Barkan, "The Witches' Sabbath: The First International Solvay Congress in Physics," in *Einstein in Context*, Mara Beller, Robert S. Cohen, and Jürgen Renn, eds. (Cambridge, UK: Cambridge University Press, 1993), 61.

17 Foucart, "At the Métropole," 2.

18 Mehra, *The Solvay Conferences on Physics*, 5.

19 Barkan, "The Witches' Sabbath," 62.

20 Walter Isaacson, *Einstein,* 168.

21 John L. Heilbron, "The First Solvay Council: 'A sort of private conference,'" in *The Theory of the Quantum World, Proceedings of the 25th Solvay Conference on Physics*, David Gross, Marc Henneaux, and Alexander Sevrin, eds. (Singapore: World Scientific Publishing, 2013), 2.

22 Paul F. State, *Historical Dictionary of Brussels* (Lanham, Md.: Rowman & Littlefield, 2015), 286.

23 Foucart, "At the Metropole," 3.

24 Heilbron, "The First Solvay Council," 2.

25 Einstein, letter to Michele Besso, December 26, 1911, *The Einstein Papers*, vol. 5, doc. 331, 241.

Chapter Five: Solvay's Search for a Universal Law

1 Ernest Solvay, sealed statement to the Belgian Academy of Sciences, July 6, 1896, quoted in H. A. Lorentz and E. Herzen, "Ernest Solvay's Views on the Relations between Energy and Mass," in *Journal of the Franklin Institute*, R. B. Owens, ed., vol. 197 (Philadelphia: Franklin Institute, 1924), 265–266.

2 Bertrams, Coupain, Homburg, *Solvay: History of a Multinational Family Firm*, 20.

3 Solvay, "The Founding," *Solvay Company History*, https://www.solvay.com/en/our-company/history.

4 William H. Nichols, "Ernest Solvay—An Appreciation," *Journal of Industrial and Engineering Chemistry* 14, no. 12 (December 1922): 1157, https://pubs.acs.org/doi/abs/10.1021/ie50156a033 .

5 Ibid.

6 L. E. Scriven, "When Chemical Reactors Were Admitted and Earlier Roots of Chemical Engineering," presentation given as part of "Chemical Engineering—A Centennial Celebration," Ohio State University, 2003, http://cbe.osu.edu/sites/default/files/uploads/scriven.pdf.

7 L. E. Scriven, "On the Emergence and Evolution of Chemical Engineering," in *Perspectives in Chemical Engineering*, Clark K. Colton, ed. (Boston: Harcourt Brace Jovanovich, 1991), 6.

8 Bertrams, Coupain, Homburg, *Solvay: History of a Multinational Family Firm*, 28.
9 Kenneth Bertrams, *A Company in History: Solvay* (Cambridge, UK: Cambridge University Press, 2013), 13.
10 Bertrams, Coupain, Homburg, *Solvay: History of a Multinational Family Firm*, 117.
11 Ibid., 99.
12 Ibid., 108–113.
13 Nichols, "Ernest Solvay," 1158.
14 Emile Tassel, *Notes on the Work Carried Out by Ernest Solvay from 1857–1914* (Brussels: Imprimerie G. Bothy, 1920), 30, quoted in Bertrams, Coupain, Homburg, *Solvay: History of a Multinational Family Firm*, 138.
15 Nicolas Coupain, "Ernest Solvay's Scientific Networks From Personal Research to Academic Patronage," paper presented at the workshop "The Early Solvay Councils and the Advent of the Quantum Era," Brussels, October 14, 2011, 10; Solvay Corporate Heritage Manager Nicolas Coupain email to author on May 20, 2020.
16 Ernest Solvay, *Notes, Letters and Speeches of Ernest Solvay* (Brussels: Lamertin, 1929), 429–430, quoted in Devriese and Wallenborn, "Ernest Solvay: The System, Law and Council," *The Solvay Councils and Birth of Modern Physics*, Marage and Wallenborn, eds., 5.
17 Bertrams, Coupain, Homburg, *Solvay: History of a Multinational Family Firm*, 100.
18 Coupain, "Ernest Solvay's Scientific Networks," 10.
19 Bertrams, Coupain, Homburg, *Solvay: History of a Multinational Family Firm*, 101.
20 Wilhelm Ostwald, *Wilhelm Ostwald, The Autobiography*, Robert Smail Jack and Fritz Scholz, eds., Robert Jack, trans. (Cham, Switzerland: Springer International, 2017), 550.
21 Ibid.
22 Ibid., 550–551.
23 Bertrams, Coupain, Homburg, *Solvay: History of a Multinational Family Firm*, 135.
24 Devriese and Wallenborn, "Ernest Solvay: The System, Law and Council," *The Solvay Councils and Birth of Modern Physics*, Marage and Wallenborn, eds., 3.
25 Solvay, "Science: Ernest Solvay's 'Fifth Child,'" *Solvay Company History*, https://www.solvay.com/en/our-company/history/1885-1914.
26 Ernest Solvay, "The Part Played by Electricity in the Phenomena of Animal Life," an address delivered on December 14, 1893, J. W. Mallet, trans. (New

York: Herman Bartsch, 1896), 57, https://collections.nlm.nih.gov/catalog/nlm:nlmuid-61010970R-bk.

27 Devriese and Wallenborn, "Ernest Solvay: The System, Law and Council," *The Solvay Councils and Birth of Modern Physics*, Marage and Wallenborn, eds., 9–10.

28 Bertrams, Coupain, Homburg, *Solvay: History of a Multinational Family Firm*, 137.

29 Ibid.

30 Édouard Herzen, "Extract from a Press Communiqué," *Solvay International Institute of Physics and Chemistry*, October 26, 1911, quoted in Devriese and Wallenborn, "Ernest Solvay: The System, Law and Council," *The Solvay Councils and Birth of Modern Physics*, Marage and Wallenborn, eds., 15.

31 Heilbron, "The First Solvay Council," in *The Theory of the Quantum World*, Gross, Henneaux, and Sevrin, eds., 9.

32 Ibid.

33 Coupain, "Ernest Solvay's Scientific Networks," 16.

34 Ibid., 19.

Chapter Six: Einstein's Enigma

1 Thomas S. Kuhn, *The Copernican Revolution: Planetary Astronomy in the Development of Western Thought* (Cambridge, Mass.: Harvard University Press, 1957), 2.

2 Isaccson, *Einstein*, p. 8-11

3 Albert Einstein, "Notes for an Autobiography," 10.

4 Isaacson, *Einstein*, 14.

5 Ibid., 37.

6 Einstein, "Notes for an Autobiography," 10.

7 Mileva Marić, *In Albert's Shadow: The Life and Letters of Mileva Marić, Einstein's First Wife*, Milan Popovic, ed. (Baltimore: Johns Hopkins University Press, 2003), 4.

8 Pauline Gagnon, "The Forgotten Life of Einstein's First Wife," *Scientific American* (December 19, 2016), http://blogs.scientificamerican.com/guest-blog/the-forgotten-life-of-einstein's-first-wife.

9 Marić, letter to Einstein, after October 20, 1897, *The Einstein Papers*, vol. 1, doc. 36, 34.

10 Einstein, letter to Mileva Marić, August 20, 1900, *The Einstein Papers*, vol. 1, doc. 73, 146.

11 Einstein, letter to Mileva Marić, October 3, 1900, *The Einstein Papers*, vol. 1, doc. 79, 152.

12 Einstein, letter to Mileva Marić, March 23, 1901, *The Einstein Papers*, vol. 1, doc. 93, 159.

13 Einstein, letter to Mileva Marić, March 27, 1901, *The Einstein Papers*, vol. 1, doc. 94, 161.
14 Pauline Gagnon, "The Forgotten Life of Einstein's First Wife."
15 Ibid.
16 Alberto Martinez, "Arguing about Einstein's Wife," *Physics World* (April 10, 2004), https://physicsworld.com/a/arguing-about-einsteins-wife/.
17 Isaacson, *Einstein*, 78.
18 Einstein, letter to Conrad Habicht, May 18 or 25, 1905, *The Einstein Papers*, vol. 5, doc. 27, 20.
19 Albert Einstein, *Einstein's Miraculous Year: Five Papers That Changed the Face of Physics*, John Stachel, ed. (Princeton, N.J.: Princeton University Press, 1998), 168.
20 Einstein, *Einstein's Miraculous Year*, 178.
21 David Cassidy, "Einstein on the Photoelectric Effect," *American Institute of Physics*, https://history.aip.org/history/exhibits/einstein/essay-photoelectric.htm.
22 Einstein, *Einstein's Miraculous Year*, 178.
23 Ibid., 174.
24 American Physical Society, "Einstein and Brownian Motion," *APS News* 14, no. 2 (February 2005): https://www.aps.org/publications/apsnews/200502/history.cfm.
25 David Cassidy, "Einstein on Brownian Motion," *American Institute of Physics*, https://history.aip.org/history/exhibits/einstein/essay-brownian.htm.
26 Einstein, *Einstein's Miraculous Year*, 98.
27 Ibid., 82.
28 Dan Falk, "What Is Relativity? Einstein's Mind-Bending Theory Explained," *NBC News* (April 13, 2018), https://www.nbcnews.com/mach/science/what-relativity-einstein-s-mind-bending-theory-explained-ncna865496.
29 Einstein, letter to Conrad Habicht, June 30–September 22, 1905, *The Einstein Papers*, vol. 5, doc. 28, 20.
30 Isaacson, *Einstein*, 483.

Chapter Seven: The First Solvay Conference: Assembly of Genius

1 Charles de Saint Sauveur, "In 1911, Paris Was Already Suffocating under the Heat Wave," *Le Parisien*, June 29, 2019, https://www.leparisien.fr/societe/juin-2019/en-1911-paris.
2 Dorothy Hoobler and Thomas Hoobler, *The Crimes of Paris: A True Story of Murder, Theft, and Detection* (New York: Little, Brown and Company, 2009), introduction.
3 R. A. Scotti, *Vanished Smile: The Mysterious Theft of Mona Lisa* (New York: Alfred A. Knopf, 2009), 14.
4 Hoobler and Hoobler, *The Crimes of Paris*, chapter 2.

5 Scotti, *Vanished Smile*, 40.
6 Quinn, *Marie Curie*, 302.
7 Fernand Hauser, "A Story of Love: Madame Curie and Professor Langevin," *Le Journal*, November 4, 1911, quoted in ibid., 304.
8 Bernadette Bensaude-Vincent, "Paul Langevin and the French Scientists at the Solvay Conferences," *The Solvay Councils and Birth of Modern Physics*, Marage and Wallenborn, eds., 38.
9 Mehra, *The Solvay Conferences on Physics*, 9.
10 Ibid.
11 Ibid., 10.
12 G. L. de Haas-Lorentz, "Reminiscences (continued)," in *H. A. Lorentz, Impressions of His Life and Work*, G. L. de Haas-Lorentz, ed., Joh. C. Fagginger Auer, trans. (Amsterdam: North-Holland Publishing, 1957), 108.
13 Elisabeth Crawford, "The Solvay Councils and the Nobel Institution," *The Solvay Councils and Birth of Modern Physics*, Marage and Wallenborn, eds., 53.
14 Barkan, *Walther Nernst*, 26.
15 Philipp Frank, *Einstein. His Life and Times*, George Rosen, trans., Shuichi Kusaka, ed. (New York: A. Knopf, 1947), 106.
16 Jeff Hughes, "Rutherford, the Cavendish Laboratory, and the Solvay Councils," *The Solvay Councils and Birth of Modern Physics*, Marage and Wallenborn, eds., 27.
17 Sarah Dry, *Curie* (London: Haus Publishing, 2003), 119.
18 Bertram Boltwood, "The International Congress of Radiology and Electricity, Brussels, Sept. 13–15, 1910," *Science* 32, no. 831 (1910): 788, https://science.sciencemag.org/content/32/831/788.
19 Foucart, "At the Métropole," 4.
20 E. Rutherford, "Conference on the Theory of Radiation," *Nature* 88, no. 2194 (November 16, 1911): 83, https://www.nature.com/articles/088082a0.
21 P. Marage and G. Wallenborn, "The First Solvay Council," *The Solvay Councils and Birth of Modern Physics*, Marage and Wallenborn, eds., 95.
22 Ibid., 96.
23 Heilbron, "The First Solvay Council," in *The Theory of the Quantum World*, Gross, Henneaux, and Sevrin, eds., 9.
24 Mehra, *The Solvay Conferences on Physics*, 24.
25 P. Langevin and M. de Broglie, eds., *The Theory of Radiation and the Quantum* (Paris: Gauthier-Villers, 1912), 77, quoted in Marage and Wallenborn, "The First Solvay Council," 101.
26 A. Eucken, ed., *The Theory of Radiation and the Quantum* (Halle a.S.: Wilhelm Knapp, 1914), 59, quoted in Barkan, "The Witches' Sabbath," 76.

27 Ibid.

28 Barkan, *Walther Nernst*, 200.

29 P. Langevin and M. de Broglie, eds., *Radiation and the Quantum*, 436, quoted in Marage and Wallenborn, "The First Solvay Council," 107.

30 Quinn, *Marie Curie*, 302.

31 Reid, *Marie Curie*, 144.

32 Ibid., 167.

33 Barkan, *Walther Nernst*, 192.

34 Kurt Mendelssohn, *The World of Walther Nernst: The Rise and Fall of German Science* (London: The Macmillan Press, 1973), 49–50.

35 E. Curie, *Madame Curie*, 284.

36 M. Brillouin, "Mr. H. A. Lorentz in France and Belgium, Some Memories," *Dutch Journal of Physics*, January 1926, quoted in G. L. de Haas-Lorentz, "Reminiscences (continued)," in *H. A. Lorentz, Impressions of His Life and Work*, G. L. de Haas-Lorentz, ed., Joh. C. Fagginger Auer, trans., 109.

37 Elisabeth Crawford, "The Secrecy of Nobel Prize Selections in the Sciences and Its Effect on Documentation and Research," *Proceedings of the American Philosophical Society* 134, no. 4 (December 1990): 412–413, http://jstor.org/stable/986896.

38 Evelyn Sharp, *Hertha Ayrton, 1854-1923: A Memoir* (London: Edward Arnold, 1926), 117, quoted in Helena M. Pycior, "Reaping the Benefits of Collaboration While Avoiding Its Pitfalls: Marie Curie's Rise to Scientific Prominence," 303.

39 Nanny Fröman, "Marie and Pierre Curie and the Discovery of Polonium and Radium," Nancy Marshall-Lundén, trans., *Nobel Media*, http://nobelprize.org/prizes/uncategorized/marie-and-pierre-curie-and-the-discovery-of-polonium-and-radium.

40 Heilbron, "The First Solvay Council," in *The Theory of the Quantum World*, Gross, Henneaux, and Sevrin, eds., 10.

41 Einstein, letter to Heinrich Zangger, November 15, 1911, *The Einstein Papers*, vol. 5, doc. 305, 222.

42 Marage & Wallenborn, "The First Solvay Council," 108.

43 P. Langevin and M. de Broglie, eds., *Radiation and the Quantum*, 451, quoted in ibid.

Chapter Eight: Marie Curie's Impossible Dream

1 Paul Langevin, "The Relations of Physics of Electrons to Other Branches of Science," in *Congress of Arts and Sciences, Universal Exposition, St. Louis, 1904*, Howard J. Rogers, ed., vol. IV (Boston: Houghton, Mifflin and Co., 1906), 121.

2 Julien Bok and Catherine Kounelis, "Paul Langevin (1872–1946) From Montmartre to the Panthéon: The Paris Journey of an Exceptional Physicist,"

EuroPhysics News 38, no. 1 (2007): 20, http://europhysicsnews.org/pdf/epn07101.

3 Quinn, *Marie Curie*, 248–249.

4 Ibid., 258.

5 Goldsmith, *Obsessive Genius*, 167.

6 Quinn, *Marie Curie*, 298.

7 Reid, *Marie Curie*, 152.

8 All quotes from Marcel Brillouin's journal and letters contained in this chapter are derived from excerpts of the papers of Marcel Brillouin as part of the collected papers of physicist Léon Brillouin, Marcel's son. Marcel's papers are located in the Papers of Léon Brillouin, 1877-1972, American Institute of Physics, College Park, Maryland, 20740. They are located in the Léon Brillouin Papers Collection, Series II, The Papers of Marcel Brillouin, 1887-1943, Box 1, Folder 6. The original papers are handwritten in French, and were translated for me by Jean-Matthieu Hautenauve and Catherine Hautenauve. In this chapter, further reference to them is made using the abbreviation MBP/AIP (Marcel Brillouin Papers/American Institute of Physics).

9 MBP/AIP, journal entry, March 11, 1911.

10 Ibid., March 29, 1911.

11 Ibid.

12 Goldsmith, *Obsessive Genius*, 167.

13 Gustave Téry, *L'Œuvre*, November 23, 1911, quoted in Quinn, *Marie Curie*, 262.

14 MBP/AIP, journal entry, March 30, 1911.

15 Ibid.

16 Ibid., April 1, 1911.

17 MBP/AIP, Marcel Brillouin letter to Jean Perrin, April 12, 1911.

18 MBP/AIP, Jean Perrin letter to Marcel Brillouin, April 17, 1911.

19 MBP/AIP, journal entry, April 26, 1911.

20 Ibid.

21 Ibid.

22 Quinn, *Marie Curie*, 263.

23 MBP/AIP, journal entry, May 6–July 21, 1911.

24 Quinn, *Marie Curie*, 296; Goldsmith, *Obsessive Genius*, 167.

25 Reid, *Marie Curie*, 166.

26 MBP/AIP, Lorentz letter to Brillouin, January 24, 1912.

27 MBP/AIP, Brillouin letter to Lorentz, late January/early Feb 1912.

28 Ibid.

29 Ibid.

30 Ibid.

31 Ibid.

32 MBP/AIP, Lorentz letter to Brillouin, February 10, 1912.

Chapter Nine: Action and Reaction

1 David A. Norris, "Krakatoa Eruption of 1883," *History Magazine,* June/July 2019, 9.

2 U.S. Geological Survey, "Volcanoes Can Affect the Earth's Climate," *Volcano Hazards Program*, https://www.usgs.gov/natural-hazards/volcano-hazards/volcanoes-can-affect-climate.

3 Norris, "Krakatoa," 11.

4 Elisabeth Crawford, *The Beginnings of the Nobel Institution: The Science Prizes, 1901–1915*, paperback edition (New York: Cambridge University Press, 1987), 111.

5 Steve Graham, "John Tyndall (1820–1893),"*Earth Observatory*, October 8, 1999, https://earthobservatory.nasa.gov/features/Tyndall.

6 Roland Jackson, "Eunice Foote, John Tyndall, and a Question of Priority," *The Royal Society Publishing* (February 13, 2019), http://doi.org/10.1098/rsnr.2018.0066.

7 American Institute of Physics, "The Carbon Dioxide Greenhouse Effect," *The Discovery of Global Warming*, January 2020, https://history.aip.org/climate/co2.htm.

8 Ibid.

9 Ibid.

10 Crawford, *Nobel Institution*, 127–128.

11 Wirtén, *Making Marie Curie*, 60–61.

12 Jean K. Chalaby, "Twenty Years of Contrast: The French and British Press during the InterWar Period," *European Journal of Sociology/Archives* 37, no. 1 (1996): 145, http://jstor.org/stable/23999515.

13 Raphael Levy, "The Daily Press in France," *Modern Language Journal* 13, no. 4 (January 1929): 297–300, http://jstor.org/stable/315897.

14 Quinn, *Marie Curie*, 309.

15 Ibid., 314–315.

16 Reid, *Marie Curie*, 172–173.

17 Shelley Emling, *Marie Curie and Her Daughters* (New York: St. Martin's Press, 2012), 7.

18 Wirtén, *Making Marie Curie*, 67.

19 Einstein, letter to Marie Curie, November 23, 1911, *The Einstein Papers*, vol. 8, docs. for vol. 5, doc. 312a, 6.

20 Goldsmith, *Obsessive Genius*, 176.

21 Mark Twain, "Chapters from My Autobiography XXII," *North American Review* 186, no. 622 (September 1907): 13, http://jstor.org/stable/25105977.

22 "Journalist Wounded in Duel," *San Francisco Call*, November 25, 1911, 12, https://chroniclingamerica.loc.gov/lccn/sn85066387/1911-11-25/ed-1/seq-12/.
23 "1911 Epee Duel: Pierre Mortier vs Gustave Tery," YouTube, https://www.youtube.com/watch?v=rh4VuCYeHtE.
24 "1911 Epee Duel: Henri Chervet vs Leon Daudet," YouTube, https://www.youtube.com/watch?v=RZQormJLOVM.
25 Edouard Drumont, *Le Temps,* April 23, 1886, quoted in Robert A. Nye, "Medicine and Science as Masculine 'Fields of Honor,'" *Osiris* 12 (December 1997): 70, http://jstor.org/stable/301899.
26 Nye, "'Fields of Honor,'" 71.
27 Wirtén, *Making Marie Curie*, 71.
28 Crawford, *Nobel Institution*, 200.
29 Svante Arrhenius, letter to Marie Curie, December 1, 1911, Archive Institute Mittag-Leffler, quoted in Karin Blanc, "The Curie Couple and the Nobel Prizes," *OpenEdition Journals*, http://journals.openedition.org/bibnum/1172.
30 Ibid.
31 Marie Curie, letter to Svante Arrhenius December 5, 1911, Center for the History of Science, Royak Swedish Academy of Sciences, quoted in ibid.
32 Hélène Ångström, letter to Marie Curie, December 9, 1911, NAF 18443 f. 154, Curie Collection, Manuscripts Department, National Library of France, quoted in ibid.
33 Marie Curie, "Radium and the New Concepts in Chemistry," *Nobel Lecture*, December 11, 1911, http://nobelprize.org/prizes/chemistry/1911/marie-curie/lecture.
34 Ibid.

Chapter Ten: A Different Dimension

1 Hertha Ayerton, letter to *Westminister Gazette*, March 14, 1909, quoted in Mason, "Hertha Ayrton," 210.
2 E. Curie, *Madame Curie*, 281.
3 Barbara W. Tuchman, *The Guns of August* (New York: Ballantine Books, 1962), 22.
4 Ibid., p. 24.
5 Robert B. Bruce, "To the Last Limits of Their Strength: The French Army and the Logistics of Attrition at the Battle of Verdun, February 21–December 18, 1916," *Army History* no. 45, (Summer 1998): 10, http://jstor.org/stable/26304799.
6 Rene Van Tiggelen, *Radiology in a Trench Coat: Military Radiology on the Western Front during the Great War*, Jan Dirckx, trans. (Brussels: Academia Press, 2013), 196.
7 M. Curie, *Pierre Curie*, 210.
8 Ibid.

9 Timothy J. Jorgensen, "How Marie Curie Brought X-Ray Machines to the Battlefield," *Smithsonian Magazine*, October 11, 2017, https://www.smithsonianmag.com/history/how-marie-curie-brought-x-ray-machines-to-battlefield-180965240/.
10 W. W. Keen, "Military Surgery in 1861 and in 1918," *Annals of the American Academy of Political and Social Science* 80, Rehabilitation of the Wounded (November 1918): 16–17, http://jstor.org/stable/1013902.
11 Ibid., 17.
12 Van Tiggelen, *Radiology in a Trench Coat*, 196.
13 Quinn, *Marie Curie*, 362.
14 E. Curie, *Madame Curie*, 297.
15 Van Tiggelen, *Radiology in a Trench Coat*, 115.
16 Ibid.
17 D. S. L. Cardwell, "Science and World War I," *Proceedings of the Royal Society. Series A, Mathematical and Physical Sciences* 342, no. 1631 (April 15, 1975): 448, http://jstor.org/stable/78744.
18 Crawford, *Beginnings of the Nobel Institution*, 192.
19 "Manifesto of the Ninety-Three German Intellectuals," *Brigham Young University Library World War I Document Archive* (October 1914), https://wwi.lib.byu.edu/index.php/Manifesto_of_the_Ninety-Three_German_Intellectuals.
20 Ibid.
21 Matthew Stanley, *Einstein's War, How Relativity Triumphed amid the Vicious Nationalism of World War I* (New York: Dutton, 2019), 75.
22 Isaacson, *Einstein*, 180.
23 Einstein, document "Manifesto to the Europeans," mid-October 1914, *The Einstein Papers*, vol. 6, doc. 8, 28.
24 Einstein, Memorandum to Mileva Einstein-Marić, July 18, 1914, *The Einstein Papers*, vol. 8, doc. 22, 32.
25 Stanley, *Einstein's War*, 58.
26 Ibid., 83.

Chapter Eleven: After the War

1 Goldsmith, *Obsessive Genius*, 192–193.
2 Walter Isaacson, "How Einstein Divided America's Jews," *The Atlantic*, December 2009, https://www.theatlantic.com/magazine/archive/2009/12/how-einstein-divided-americas-jews/307763/.
3 E. Curie, *Madame Curie*, 329.
4 Isaacson, *Einstein*, 300.
5 Einstein, *The World As I See It*, 39.
6 Ibid., 41–42.

7 Associated Press, "Madame Curie Enchanted at Warm Heart of West; Talks of Arizona Canyon Tour," *Weekly Prescott Journal-Miner*, June 15, 1921, 1, https://azmemory.azlibrary.gov/digital/collection/sn85032923/id/5144/.

8 Ibid.

9 Marie Curie, "Impressions of America," undated typescript, Box 3, Folder 21, Series II: Catalogued Manuscripts, Marie Mattingly Meloney Collection on Marie Curie, Rare Book and Manuscript Library, Butler Library, Columbia University (New York, N.Y.).

10 "Degrees Conferred on Madame Curie," *Associated Press to Gazette-Times*, May 24, 1921, 1, https://news.google.com/newspapers/the-gazette-times /may-1921.

11 "Pitt Degree Conferred on Mme. Curie," *Gazette-Times*, May 27, 1921, 2, https://news.google.com/newspapers/the-gazette-times/may-1921.

12 Joel O. Lubenau and Edward R. Landa, *Radium City: A History of America's First Nuclear Industry* (Pittsburgh: Senator John Heinz History Center, 2019), 91, https://www.heinzhistorycenter.org/magazine/Radium -City.pdf.

13 Marie Curie, "Impressions of America," undated memorandum of Madame Curie for the press, Box 3, Folder 25, Series II: Catalogued Manuscripts, Marie Mattingly Meloney Collection on Marie Curie, Rare Book and Manuscript Library, Butler Library, Columbia University (New York, N.Y.).

14 Ibid.

15 United Nations, "League of Nations: Intellectual Cooperation," *United Nations Archives Geneva Research Guides*, https://libraryresources.unog.ch /lonintellectualcooperation/ICIC.

16 E. Curie, *Madame Curie*, 339.

17 Stanley W. Pycior, "Marie Sklodowska Curie and Albert Einstein, a Professional and Personal Relationship," *Polish Review* 44, no. 2 (1999): 138, http://jstor.org/stable/25779116.

18 E. Curie, *Madame Curie*, 284.

19 Stanley W. Pycior, "Curie and Einstein," 134.

20 Roger Highfield and Paul Carter, *The Private Lives of Albert Einstein* (New York: St. Martin's Press, 1993), 157.

21 Einstein, letter to Elsa Lowenthal, August 11?, 1913, *The Einstein Papers*, vol. 5, doc. 465, 347.

22 Hans C. Ohanian, *Einstein's Mistakes: The Human Failings of Genius* (New York: W. W. Norton & Company, 2008), xii.

23 Isaacson, *Einstein*, 543.

Epilogue: Follow the Science

1 Mary Jo Nye, "N-Rays: An Episode in the History and Psychology of Science," *Historical Studies in the Physical Sciences* 11, no. 1 (1980): 129, http://jstor.org/stable/27757473.

2 Ibid., 130.

3 Ibid., 132.

4 American Physics Society, "September 1904: Robert Wood Debunks N-rays," *APS News* 16, no. 8 (August/September 2007), http://aps.org/aps-news /august-september-2007/robert-wood-debunks-n-rays.

5 R. W. Wood, "The n-Rays," *Nature* 70, no. 1822 (September 29, 1904): 530, http://zenodo.org/nature/29-september-1904/the-n-rays.

6 American Physics Society, "September 1904: Robert Wood."

7 Wood, "The n-Rays," 531.

8 Nye, "N-rays," 149.

9 Isaacson, *Einstein*, 365.

10 Goldsmith, *Obsessive Genius*, 219.

11 Ibid., 230.

12 Albert Einstein, *Ideas and Opinions*, Sonja Burgmann, trans. (New York: Three Rivers Press, 1954), quoted in Pycior, "Curie and Einstein," 141.

13 Nobel Prize, "Nobel Prize Awarded Women," *Nobel Media*, http://nobelprize .org/prizes/lists/nobel-prize-awarded-women (accessed October 13, 2020).

14 Ibid.

15 National Science Foundation, "Women and Minorities in the S&E Workforce," *Science and Engineering Indicators 2014*, chapter 3, https://www .nsf.gov/statistics/seind14/index.cfm/chapter-3/c3h.htm.

16 Gerald Holton, "On the Art of Scientific Imagination," *Daedalus* 125, no. 2 (Spring 1996): 202, http://jstor.org/stable/20013446.

17 Ibid.

18 Nicolas Coupain, "Ernest Solvay—Mundaneum," Google Arts & Culture, slides 31–32, https://artsandculture.google.com/exhibit/QQ8X_Kko.

19 International Solvay Institutes, "Welcome to the Solvay Institutes," http://solvayinstitutes.be.

20 Jean-Marie Solvay, Internet conversation with author, September 7, 2020.

21 Ibid.

22 International Solvay Institutes, "The Chemistry of the Future Is the Future of Mankind," *Echo Connect*, https://solvay.lecho.be/en/.

23 J.-M. Solvay, conversation with author.

24 Isaac Newton, letter to Robert Hooke, February 1675, Box 12/11, Folder 37, Simon Gratz autograph collection (#0250A), Historical Society of Pennsylvania (Philadelphia), https://digitallibrary.hsp.org/index.php/Detail/objects/9792.

INDEX

D